Jet-Induced Effects
The Aerodynamics of Jet- and Fan-Powered V/STOL Aircraft in Hover and Transition

Jet-Induced Effects
The Aerodynamics of Jet- and Fan-Powered V/STOL Aircraft in Hover and Transition

Richard E. Kuhn
NACA/NASA retired

Richard J. Margason
Lockheed Martin Co., Inc.

Peter Curtis
BAE Systems

Volume 217
PROGRESS IN ASTRONAUTICS AND AERONAUTICS

Frank K. Lu, Editor-in-Chief
University of Texas at Arlington
Arlington, Texas

Published by the
American Institute of Aeronautics and Astronautics, Inc.
1801 Alexander Bell Drive, Reston, Virginia 20191-4344

1-56347-841-2

Foreword

THIS TEXT on the most crucial aspect of V/STOL aircraft is co-authored by three experts in this field whose combined experience stretches many decades. The authors have produced a timely volume in view of the development of the F-35B STOVL version of the Joint Strike Fighter and various proposals for V/STOL unmanned aircraft. The volume draws on the authors' vast experience, covering the salient features that affect the flight of V/STOL aircraft during hover and transition to horizontal flight. The authors have identified important technological challenges and have outlined methods that were applied to address them. The text is written in a clear style and it provides methods for estimating the complex aircraft/flowfield interactions through empirical correlations. Thus, the methodology applied is easy to understand and allows the reader to quickly grasp the material. Moreover, the authors share their personal, first-hand insights which are of tremendous value in understanding the complex flows that arise in the V/STOL flight regime.

Frank K. Lu
May 2005
Editor-in-Chief

Table of Contents

Preface

OVER THE past half century there have been many experimental investigations of jet and fan lift VTOL aircraft configurations and of the complex interaction between propulsion and aerodynamics. In the last twenty years there have also been more ambitious computational investigations using contemporary computational fluid dynamic (CFD) methods. The authors have been closely involved in much of this research and the related studies or have been close to many of those who made the studies. The material on which this book is based came from hundreds of reports and papers.

The purpose of this book is to gather into one place what is currently known about the flow fields and the methods of estimating the effects caused by the interaction between propulsion and aerodynamics of the aircraft, as well as on such related effects as hot gas reingestion and the effects on the ground environment into one place. Most of the methods for estimating the effects of the flow fields are based on empirical correlations of experimental data.

The first author started the book but soon realized that he needed help in several areas, such as, the status of CFD methods, jet in a crossflow effects, hot gas ingestion and the effects on the ground environment. The coauthors covered these areas and also made many valuable contributions throughout the book.

The authors wish to express their appreciation to their many colleagues and other researchers whose reports and papers provided so much of the material and the understanding needed to produce this book. We would also like to thank our wives for their patience during the long time spent on this effort.

Richard Kuhn
Richard Margason
Peter Curtis
April 2006

Chapter 1

Introduction

OVER THE later half of the 20th century, many jet and fan-powered V/STOL (vertical/short takeoff and landing) aircraft concepts have been proposed. These vehicles depend on jet or fan thrust to provide lift for hover and to provide both thrust and lift in very low-speed flight. These thrust forces are produced by both lifting jets and control jets. As the flight speed increases from hover to wing-borne flight (transition flight), the aerodynamic forces and moments from conventional wing and tail/canard surfaces become the dominant force generators. In the transition speed regime there is a strong interaction between thrust-induced flowfields and aerodynamic-induced flowfields. The resultant combined flowfields are very complex and strongly affect the performance of V/STOL aircraft. Traditionally, experimental investigations of either powered models or flight vehicles have been used to understand and quantify these effects. Recently modern computational-fluid-dynamics (CFD) methods have become a useful tool. The purpose of this publication is to describe the aerodynamics of jet and fan-powered V/STOL aircraft in hover and transition. This includes experimental data, empirical correlations, and CFD results.

I. Operational Jet V/STOL Aircraft

Although many aircraft concepts were built, only two jet types, the Russian Yak-38 lift-plus-lift/cruise concept and the British-developed Harrier deflected thrust concept, have entered service. Currently a third operational aircraft, the Joint Strike Fighter F-35B, lift-fan-plus-lift/cruise concept, is under development to replace the Harrier. The Harrier family of subsonic tactical aircraft was developed from the late 1950s through to the present day[1,2] and has been used extensively by the U.S. Marines, the British Royal Navy and Air Force, and the military services of several other countries. They range from the prototype P-1127, the tripartite Kestrel, and the early day-attack Hawker Harrier GR.1 and AV-8A, through the radar-equipped Sea Harrier FA.2, with AIM-120 beyond visual range capability, and on to the McDonnell Douglas led AV-8B family. This aircraft has developed from day attack to night attack and has also evolved into a radar-equipped AIM-120 armed variant. Figure 1.1 shows a USMC AV-8B day-attack aircraft. The basic layout for all of these aircraft, and the key to their V/STOL success, has remained unchanged throughout this time.

Fig. 1.1 U.S. Marine Corps AV-8B Harrier V/STOL ground-support aircraft.

The Harrier is powered by a single Rolls Royce Pegasus vectored thrust engine mounted at the center of the fuselage. The Pegasus was developed through many variants during its early years, including the Pegasus 2 for the first prototypes, the Pegasus 5 for the XV-6A Kestrel demonstrator aircraft, and the Pegasus 6 for the earliest service aircraft. The major variant that has been used for most of the service aircraft is the Pegasus 11, although there have been a number of subvariants of this engine. As with the basic aircraft, the layout of the engine has remained unchanged, although thrust has increased from 11,300 to 23,500 lb. The Pegasus 11 (Fig. 1.2) is a moderate bypass ratio engine fitted with four exhaust nozzles, two on each side. The two aft nozzles are powered by the hot exhaust, and the two front nozzles are powered by air from the fan section. All four nozzles can be collectively vectored from 0 (aft) to 100 deg (forward of the vertical, downwards). This enables the same basic propulsion system to power the aircraft in cruise and in V/STOL modes, such as short takeoff, short landing and vertical landing. In hover, and at low speeds, aircraft control is augmented by high-pressure compressor bleed air fed to nozzles at the wing tips and at the nose and tail of the aircraft.

By contrast, the Yak-38 (Fig. 1.3) has three separate engines, two of which are used only in V/STOL mode. These two lift engines are mounted in tandem immediately behind the cockpit, while the third engine has its exhaust divided into twin vectoring nozzles with one nozzle on each side of the fuselage just aft of the wing. This engine provides both lift thrust and all of the aircraft's cruise thrust. The layout is referred to as a lift-plus-lift/cruise configuration. It was deployed on the Kiev class carriers in the Russian Navy.

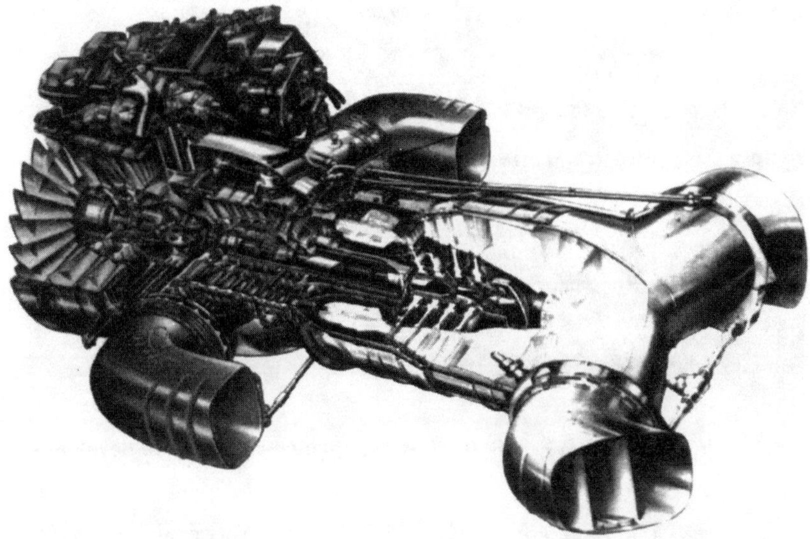

Fig. 1.2 Rolls Royce Pegasus 11 vectored thrust engine used in the Harrier.

During the 1990s, a competitive program was undertaken by the United States and United Kingdom to develop a supersonic short takeoff and vertical landing (STOVL) replacement for the Harrier. A comprehensive description of the experimental development programs was presented by McCarthy.[3] There were several oral presentations of the competing Boeing X-32B and Lockheed-Martin X-35B aircraft at the 1998 and 2000 International Powered Lift Conferences. The Boeing X-32BSTOVL[4] used two lift jets and seven

Fig. 1.3 Russian YAK-38 lift-plus-lift/cruise configuration.

Fig. 1.4 Lift fan STOVL F-35B Joint Strike Fighter currently in development.

control jets. The downselected Lockheed-Martin (Fig. 1.4) F-35B STOVL Joint Strike Fighter (JSF) uses a lift fan and deflected main engine for lift.[5–9] An extra turbine stage is added to the main engine to drive the lift fan that is mounted ahead of the engine as shown in Fig. 1.5.

A three-bearing swiveling nozzle is used to deflect the engine exhaust to the vertical when the lift fan is engaged. Differential thrust between the main engine nozzle and the lift fan provides pitch control and fan bleed air to laterally disposed nozzles in the wing provide roll control. Yaw control is provided by lateral deflection of the main engine exhaust.

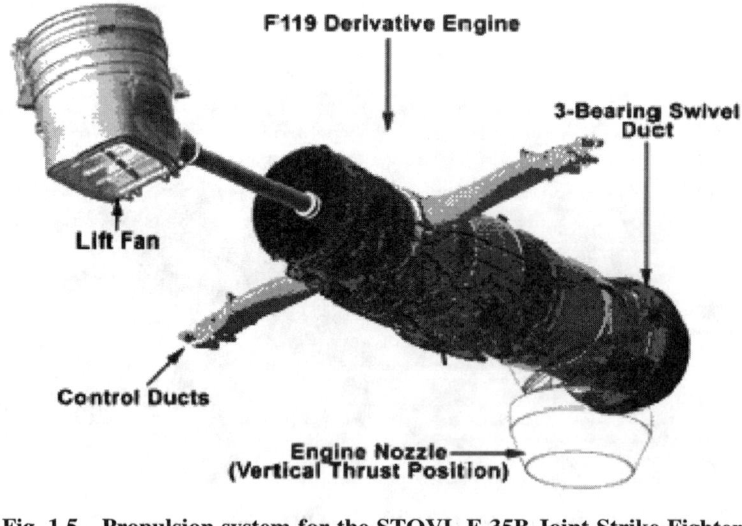

Fig. 1.5 Propulsion system for the STOVL F-35B Joint Strike Fighter.

II. Uninhabited Aerial Vehicle VTOL Aircraft

Although uninhabited aerial vehicles (UAVs) were used as early as the U.S. Civil War, the history of UAVs has been uneven. In the United States the Ryan Aeronautical Co. built the Firebee Lightning Bugs in the 1950s. There were several programs in the next 50 years. Successful modern UAV development was undertaken in Israel starting 30 years ago, and they have now become a standard military weapon. In the last several years UAVs have been a rapid growth area. This growth can be seen by comparing two overview articles from 2003[10] and 2005[11] published by AIAA in *Aerospace America* magazine. Table 1.1 summarizes the reported UAV programs.

These surveys show that there was a rapid increase in the number of current UAV aircraft programs, from 55 to 165, in less than three years. The development of VTOL UAVs has been recent and rapid, going from no programs in 2003 to 52 programs in 2005. Although most of these VTOL UAV were rotorcraft, there are some higher disk loading concepts. Although this survey identified no VTO UAVs in 2002, there was a paper in 2002[12] that described tail-sitter UAV research in Australia.

These UAVs include ducted-fan, fan-in-wing, tail sitter, and flapping-wing configurations. An example of a ducted fan[13] is the electric microaerial vehicle (E-MAV) shown in Fig. 1.6. This a lower disk loading ducted fan that takes off, hovers, and lands vertically. As a ring-wing design, it rotates to near horizontal flight for longer range or loiter missions. This is a very small 9-in.-diam battlefield vehicle for soldiers to conduct intelligence, surveillance, target acquisition, and reconnaissance missions. It is not known what additional VTOL UAV aircraft configurations will be developed in the next few years.

Another area for current development is the personal air vehicle (PAV). The PAV will require an improved or perhaps even an automated air traffic control (ATC) system to achieve needed safety standards required for the projected mass market. This robust ATC might require that PAVs become a variation of the UAV aircraft. One objective of these aircraft is to provide an alternative to the automobile for medium- and longer-range trips. PAVs are described in several recent papers.[14–18] Typically these aircraft use ducted lift fans for VTOL operation. One development effort[14] has improved rotary engines to obtain a lightweight, reliable power plant, whereas another development effort[17,18] has improved the efficiency of the fan. To date, PAVs have focused on vertical takeoff and vertical landing operation. As these efforts mature, it is

Table 1.1 UAV program summary

Year	Total no. of UAV programs to date	Current programs	VTOL UAV aircraft: Rotorcraft	Tilt wing	Ducted fan	Other concepts
2003	238	55	0	0	0	0
2005	411	165	39	4	3	6

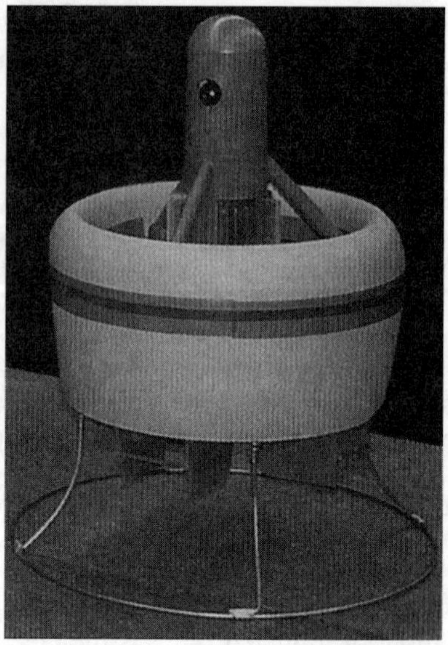

Fig. 1.6 Allied Aerospace iSTAR electric microaerial vehicle.[13]

expected that PAVs would also improve both STOL and cruise performance before they become economically viable vehicles.

III. Basic Flowfields

An appreciation of the flowfields under and around a jet V/STOL aircraft is necessary to understand the aerodynamic effects of the jet flow as well as the empirical and CFD methods used for estimating them. In transition flight the freestream deflects the jet flow, and the jet flow alters the freestream flow with profound effects of the lift and moments experienced by the configuration. In hover out-of-ground-effect (Fig. 1.7) the jet or fan streams entrain air, inducing suction pressures on the lower surface of the aircraft causing a small download or lift loss.

Close to the ground, however, the download can be considerably larger. With a single jet configuration the impinging jet flows radially outward from the impingement point, and the entrainment area is greatly increased, causing an increase in download, which varies inversely with the height of the aircraft above the ground.

With multiple jets, on the other hand, an upflow or "fountain" is created where the wall jets, flowing outward from the impingement points of adjacent jets, meet. This fountain flow produces a lifting force where it impinges on the lower surface of the aircraft, partially offsetting the download created by the entrainment action of the wall jet flow on the ground. The strength of this lifting force depends on the number and arrangements of the jets.

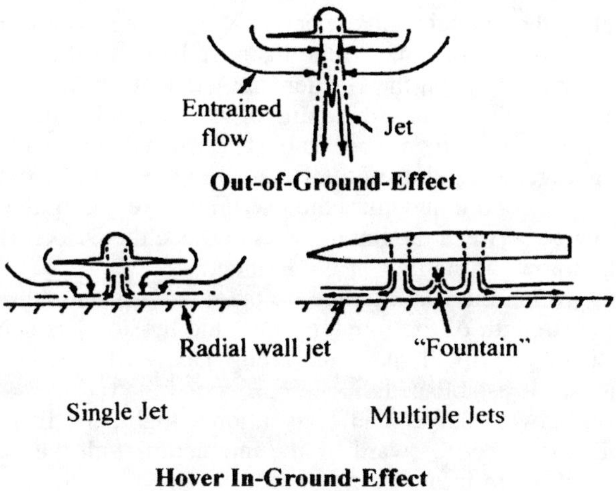

Fig. 1.7 Flowfields in hover.

The fountain between a pair of jets (Fig. 1.8) is fan shaped, originating from the stagnation line generated where the wall jets, flowing radially outward from the impingement points of each of the jets, meet.

With three jets there are three stagnation lines, and where they intersect the fountain is reinforced, and thus stronger than for two jets. With four jets the effect is even stronger, and depending on the jet spacing, can produce a net lift gain close to the ground.

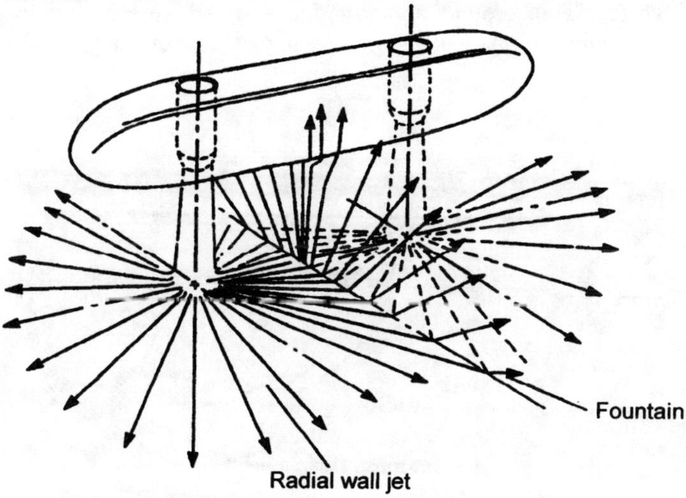

Fig. 1.8 Fountain flow formed between two vertically impinging jets.

The strength of the fountain and the height at which it is established are, to the first order, inversely proportional to the distance between the nozzles. For jets that are closely spaced, as on the Harrier, the jets tend to merge close beneath the aircraft as a result of the spreading of the jets with distance from the nozzle exit. The fountain can only be established when there is sufficient distance between the impingement points of the jets to allow the wall jet on the ground to be established. This condition defines the maximum strength of the fountain. The greater the distance between the impingement points, the weaker the ground jets at the stagnation point at the foot of the fountain.

The angle at which the jets impinge on the ground is an additional factor in determining the strength of the fountain. For side-by-side jets splaying the jets in toward each other will increase its strength but will reduce the height at which the fountain is established.

In transition between hover and conventional flight, the lifting jet, or fan, stream (Fig. 1.9) is swept rearward by the interaction with the freestream, and the jet flow is rolled up into a pair of vortices. These vortices, along with the entrainment action and blockage effect of the jet, induce suction pressures behind and beside the jet and positive (or lifting) pressures ahead of the jet. In most cases these induced pressures result in a net loss in lift and a nose-up pitching moment.

However, if the jets or fans are located near the trailing edge of a lifting surface they will induce a favorable lift through their "jet flap" action. This lift gain is usually accompanied by a nose-down pitching moment.

In addition the flow into the inlets causes a ram drag and, generally, a nose-up pitching moment. Also in crosswinds, or sideslipping flight, these inlet and exit flows can induce significant rolling and yawing moments.

All of the preceding phenomena are present but modified by the proximity of the ground during STOL operation (Fig. 1.10). In addition the wall jet sheet flowing forward on the ground is opposed by the freestream and rolled up into a ground vortex. This ground vortex induces an additional download, or lift loss, on the configuration, which is at least partially offset by a reduction in

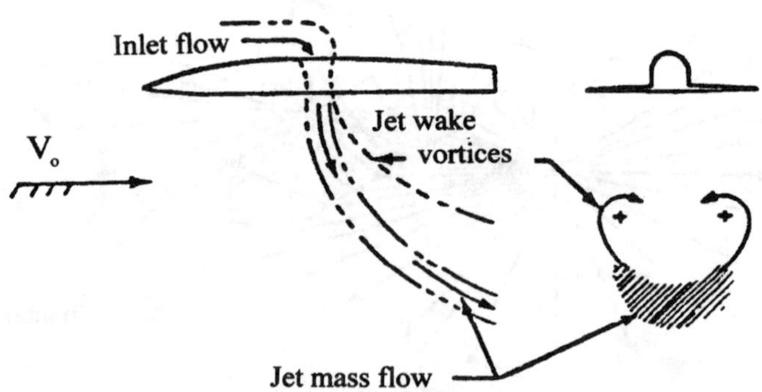

Fig. 1.9 Flowfield in transition between hover and conventional flight (out-of-ground effect).

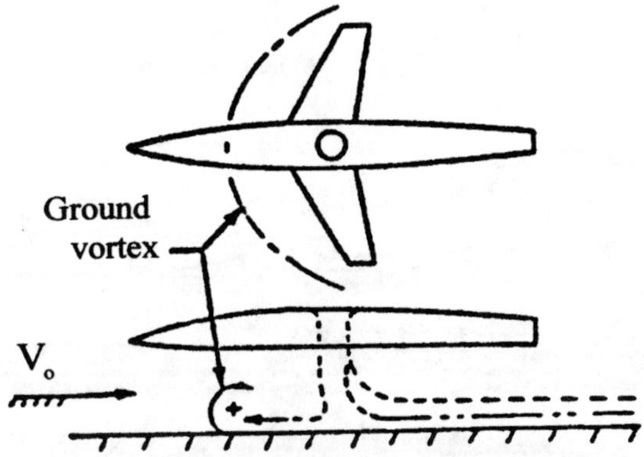

Fig. 1.10 Transition in-ground effect (STOL operation).

the wake vortex system-induced download caused by the truncation of the jet wake by the ground.

The position and strength of the ground vortex are also primary factors in determining the extent of the hot gas, dust and debris, and spray problems that can be generated in STOL operations.

IV. V/STOL Analysis Approaches

The basic flowfields just described and the development of various aircraft concepts have provided an extensive understanding of the flowfields and the operational environment of jet V/STOL aircraft. An early book by McCormack[19] provided an overview of the aerodynamics of V/STOL flight. Another book by Campbell[20] provided a review of research conducted at NASA Langley Research Center and describes many of the V/STOL aircraft that have been developed. Nelms and Anderson[21] summarized most of the nonhelicopter V/STOL concepts developed in the United States. More recently Andrews[22] described the V/STOL developments sponsored by the U.S. Navy. The French VTOL developments are described in Ref. 23.

To hover and have V/STOL capability, the vertically directed thrust from the jets or fans must exceed the weight of the aircraft. The amount by which the rated thrust of the engine must exceed the weight of the aircraft depends on engine installation losses and the aerodynamic forces induced on the aircraft by the exiting jet or fan streams. Directing the jet exhaust downward onto the ground generates the complex and multifaceted environment shown in Fig. 1.11.

V/STOL aircraft both in hover and during transition between hover and wing-borne flight have been the subject of many studies over the past 50 or more years. Most studies have been experimental investigations because, at low speeds, the flowfields involve large deflections of the local flow as well as significant amounts of viscous mixing and flow separation. Some of these studies were tests to gather design and performance data on proposed configurations, whereas other studies were specialized investigations that studied specific phenomena. These experimental investigations have identified the associated

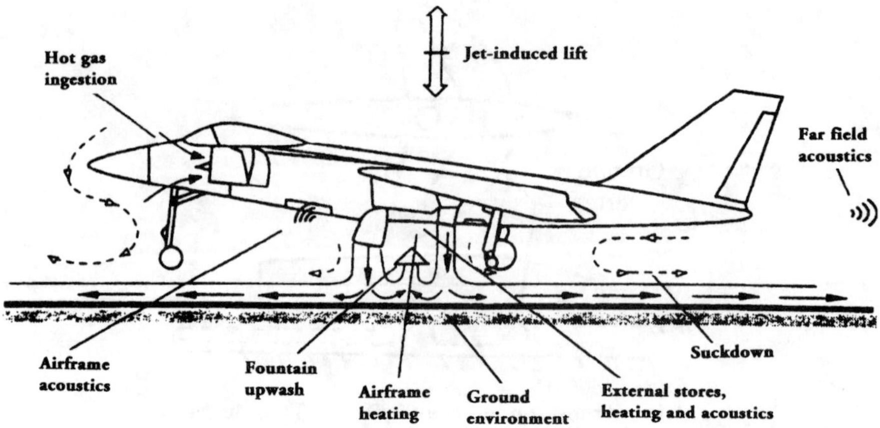

Fig. 1.11 Hovering environment for jet-powered V/STOL aircraft.

jet flow-induced effects and the resultant aerodynamic characteristics. A good understanding of most of the flow phenomena has evolved from these studies. These data were typically analyzed using plots to identify trends that demonstrate specific effects. Where it was appropriate, parameters were formed to collapse the data. Curve fits could then be used to represent these specific effects. Then a group of these data correlations and curve fits can be used to form a procedure, which can be used to describe a specific flow phenomenon.

This is the basis of empirical methods that estimate the effects of these flows on the aerodynamic characteristics of V/STOL airplanes. The following chapters of this book will review what is known about the flowfields, their effects in various flight regimes, and will present empirical methods that are available for estimating the consequent induced aerodynamic characteristics.

In addition good progress is currently being made in developing CFD methods to predict these flows and their effects. There is a discussion of the many advances in the application of CFD for the analysis of V/STOL aircraft performance. In the last 20 years CFD was first applied to modular problems such as a jet in-ground effect and a jet in a crossflow. Recently CFD has been used to represent complete powered V/STOL aircraft configurations. Some of the future trends in CFD are also discussed.

The methods presented here are intended for use only in preliminary design work and to give a general indication of the effects of the primary configuration variables. The induced effects are a complex function of many configuration variables, and the development of a V/STOL aircraft will require both CFD solutions and careful experimental investigations to accurately understand the induced forces and moments and to determine the aircraft's performance.

References

[1]Fozard, J. W., *The Jet V/STOL Harrier—An Evolutionary Revolution in Tactical Air Power*, British Aerospace Aircraft Group, Kingston-Upon-Thames, July 1978.

[2]Pryce, M. J., Hirschberg, M. J., and Farara, C., "Improving the Harrier: Projected Developments of the Pioneering V/STOL Combat Aircraft 1957–1990," Society of Automotive Engineers, Paper 2005-01-3195, Oct. 2005.

[3]McCarthy, K., "The JSF STOVL Performance Process from Small-Scale Database to Flight Test Demonstration," AIAA Paper 2002-6002, Nov. 2002.

[4]Roberts, A., "Testing the Rolls Royce Lift System for the X-32B Boeing Joint Strike Fighter," AIAA Paper 2002-5993, Nov. 2002.

[5]Buchholz, M., "Highlights of the JSF X-35 STOVL Jet Effects Test Effort," AIAA Paper 2002-5962, Nov. 2002.

[6]Gerhold, M., and Buchholz, M., "Design and Test of the X35B Hover Pit," AIAA Paper 2002-6005, Nov. 2002.

[7]Palmer, P., "Design, Fabrication & Validation of the JSF 7.5% Plenum Wing Concept for SJE Testing," AIAA Paper 2002-5964, Nov. 2002.

[8]Walker, G., "X-35B STOVL Flight Control Law Design and Flying Qualities," AIAA Paper 2002-6018, Nov. 2002.

[9]Mange, R., and Palmer, P., "An Overview of the Lockheed Martin JSF PWSC STOVL Aerodynamic Improvement Program," AIAA Paper 2002-5963, Nov. 2002.

[10]Wilson, J. R., "UAVs: A Worldwide Roundup," *Aerospace America*, Vol. 41, No. 6, June 2003, pp. 30–35.

[11]Wilson, J. R., "UAV Programs Around the World," *Aerospace America*, Vol. 43, No. 4, Sept. 2005, pp. 26–34.

[12]Stone, R., "The T-Wing Tail-Sitter Research UAV," AIAA Paper 2002-5970, Nov. 2002.

[13]Gehm, R., "Victrex PEEK Composite in A380, Portable UAV," *SAE Aerospace Engineering*, Sept. 2005, pp. 31–32.

[14]Moller, P. S., "Volantor-A Powered Lift Aircraft for Personal Use," Society of Automotive Engineers, Paper 2005-01-3183, Oct. 2005.

[15]Yoeli, R., "Ducted Fan Utility Vehicles and Other Flying Cars," AIAA Paper 2002-5995, Nov. 2002.

[16]Yoeli, R., "The Operational Potential of Manned, Ducted-Fan VTOL Vehicles," Society of Automotive Engineers, Paper 2005-01-3184, Oct. 2005.

[17]Bulaga, R., "Springtail EFV/Dragonfly UMR," Society of Automotive Engineers, Paper 2005-01-3185, Oct. 2005.

[18]Bulaga, R., "Ducted Fan Efficiency and Noise," Society of Automotive Engineers, Paper 2005-01-3186, Oct. 2005.

[19]McCormack, B. W., Jr., *Aerodynamics of V/STOL Flight*, Academic Press, New York, 1967, pp. 231–308.

[20]Campbell, J. P., *Vertical Takeoff and Landing Aircraft*, Macmillan, NewYork, 1972, pp. 106–150.

[21]Nelms, W. P., and Anderson, S. B., "V/STOL Concepts in the United States: Past, Present, and Future," AGARD-R-710, Paper 4, May–June 1984.

[22]Andrews, H., "Navy V/STOL Efforts," AIAA Paper 2002–5981, Nov. 2002.

[23]Hirschberg, M., Mueller, T., and Rocher, A., "French High-Speed V/STOL Concepts of the Twentieth Century," AIAA Paper 2002-5978, Nov. 2002.

Chapter 2

Lift Loss In Hover

THE LARGEST hovering lift losses generally occur close to the ground. However, even when the aircraft is out-of-ground effect the entrainment action of the jet(s) can induce a significant download.

The estimates of the effects of ground proximity on the suckdown in hover and the estimates of the jet-induced effects in transition, out-of-ground effect, should reduce to the out-of-ground effect level at altitude and in hover, respectively. The lift loss at altitude is therefore presented first.

I. Lift-Loss Out-of-Ground Effect

When a subsonic jet issues into still air, it entrains and mixes with the surrounding air. This entrainment of the surrounding air causes it to spread and decay as shown in Fig. 2.1. The full velocity of the exiting jet stream is maintained only in the core region. Nozzles that are driven by a plenum chamber with a large contraction ratio will have a uniform velocity profile as shown in Fig. 2.1. The core region forms a conical space with its base at the jet exit, which tapers with increasing distance from the nozzle diameter d, at the nozzle exit, to a zero radius at some distance from the nozzle exit. For a low turbulence jet the length of the core region can be up to about six diameters. For more turbulent jets, such as those typically generated by turbojets' engines, the core length decreases as the turbulence increases. The edges of the jet mix with the ambient stationary environment forming a turbulent mixing region. The velocity at the outer jet edge decreases rapidly as sketched in Fig. 2.1. Beyond the core region the maximum jet velocity decays with increasing distance. The jet flow forms a velocity profile similar to that described by an error function curve.

The decay of jet dynamic pressure in the turbulent mixing region is given[1] by

$$\left(\frac{q_z}{q_e}\right)^{(1/2)} = \frac{k}{z/d} \tag{2.1}$$

where k has a value between 6 and 6.5 for well-formed subsonic jets.

The suction pressures are induced on the surface area surrounding the jet by this entrainment action of the jet. These pressures are highest at the edge of the jet and decrease rapidly with distance from the jet as shown in Fig. 2.2 The secondary peak in pressures near the edge of the plate is caused by the

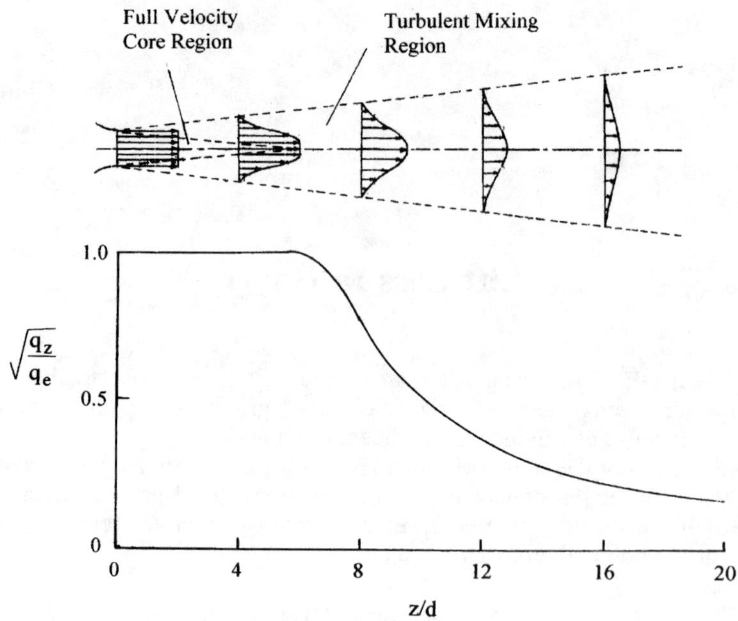

Fig. 2.1 Decay and spread of the jet efflux with distance downstream from the nozzle exit.

separation of the entrained flow at the sharp edge of the flat plate on which these pressures were taken.

Although the induced suction pressures Δp are small relative to the jet dynamic pressure q_e, they can, when integrated over the surface area surrounding

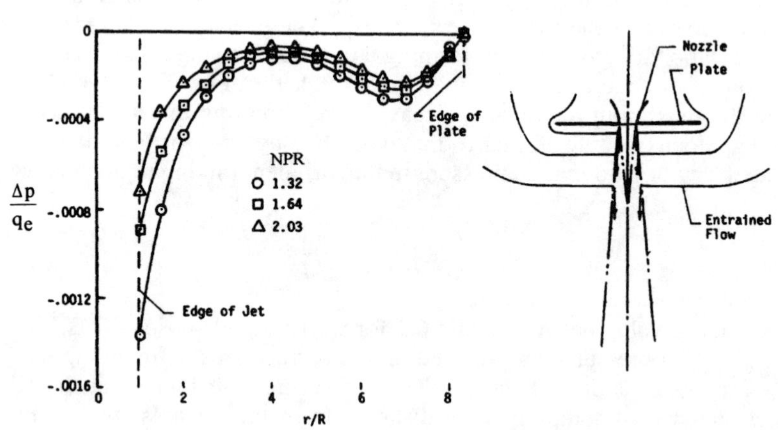

Fig. 2.2 Flowfield and pressures induced on a flat plate out-of-ground effect.[2]

the jet, add up to a significant lift loss. There have been a number of investigations of these out-of-ground-effect lift losses.[2–11] These studies have shown that the lift loss is a function of the ratio of the planform area S to the total jet exit area A_j and to the jet decay characteristics. In some cases the jet will decay more rapidly than normal, and the out-of-ground-effect lift loss will increase accordingly. These cases include configurations with other than circular nozzles, installations with a turn (usually to deflect the flow to the vertical) immediately upstream of the nozzle, and nozzles designed specifically to promote a rapid decay so as to reduce the ground surface erosion problems (see Chapter 6).

The correlation that provides the basis for estimating the lift losses induced by rapidly decaying jets[2] is shown in Fig. 2.3. The lift loss out-of-ground effect is given by

$$\frac{\Delta L_\infty}{T} = -0.009\sqrt{\frac{S}{A_j}}\mathrm{DP} \tag{2.2}$$

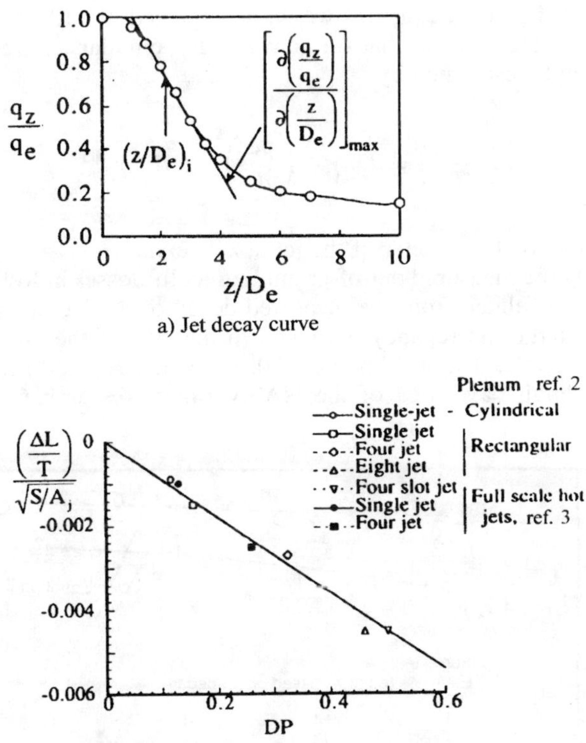

b) Lift loss correlation

Fig. 2.3 Correlation of lift loss with jet decay.

where the jet decay parameter DP is given by

$$\mathrm{DP} = \sqrt{-\left[\partial\left(\frac{q_z}{q_e}\right)\Big/\partial\left(\frac{z}{D_e}\right)\right]_{\max}\Big/\left(\frac{z_i}{D_e}\right)} \tag{2.3}$$

and, as shown in Fig. 2.3a, $[\partial(\frac{q_z}{q_e})/\partial(\frac{z}{D_e})]_{\max}$ is the maximum rate of change of the dynamic pressure q_z with distance from the jet exit and z_i is the point at which it occurs.

The correlation from Ref. 2 did not find any effects as a result of the nozzle pressure ratio (NPR). References 4 and 10 found that the decay parameter DP is inversely proportional to the square root of the NPR for NPR up to about three. The expression [Eq. (2.2)] for the lift loss out-of-ground effect can therefore be rewritten as

$$\left(\frac{\Delta L_\infty}{T}\right) = -0.012\sqrt{\frac{S}{A_j}}\,\mathrm{DP}(\mathrm{NPR})^{-0.5} \tag{2.4}$$

Unfortunately the decay characteristics of the jet are not always available. The analysis of Ref. 10 showed that, as would be expected, the lift loss is related to the perimeter of the jets and that for conventional circular and rectangular jets the lift loss can be estimated by

$$\left(\frac{\Delta L_\infty}{T}\right) = K_\infty\sqrt{\frac{S}{A_j}}\left(\frac{\mathrm{per}}{d_e}\right)^{1.58}(\mathrm{NPR})^{-0.5} \tag{2.5}$$

where per is the total perimeter of the jet nozzle exits.

Surprisingly the measured out-of-ground-effect lift losses in Ref. 10 were less than half that calculated from the measured decay of the jets used as shown in Fig. 2.4. This large discrepancy is related to the size of the "room" in which the tests were made. The tests of Ref. 10, on which Eq. (2.5) is based, were made in the "high bay" area of the NASA Ames Research Center 40 × 80

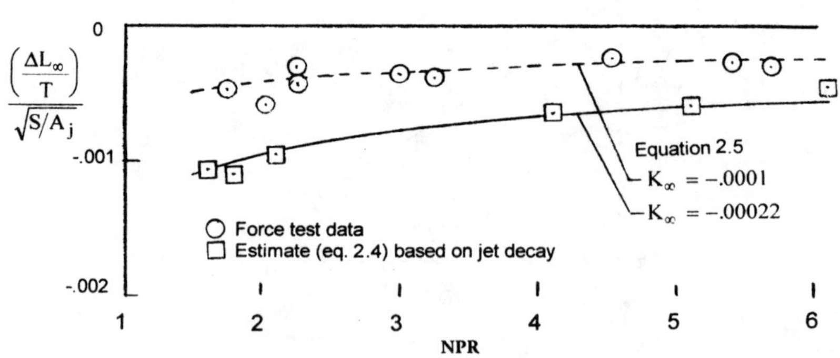

Fig. 2.4 Effect of NPR on the out-of-ground-effect lift loss for conventional jets.[10]

Foot Wind Tunnel, which had dimensions of about 1000 times the diameter of the jets used. The earlier work, on which Eq. (2.2) is based, was done in a room with dimensions of 200 to 300 times the diameter of the nozzles used. That is, the tests in the high bay area were made in a room 30 to 100 times the volume of the earlier work.

It is probable that there were undetected recirculating currents present during the earlier work that led to the higher lift loss out-of-ground effect, in which case the lower levels shown in Fig. 2.4 represent the true out-of-ground-effect lift loss. However the estimate of the out-of-ground-effect lift loss is only one part of the total estimate of jet-induced effects.

Inasmuch as the data on which the methods for estimating the ground effects were taken in the same rooms as the lift loss out-of-ground effect discussed here, a value of $K_\infty = -0.00022$ should be used with Eq. (2.5) in making the total ground effects estimates.

Equation 2.5 applies for conventional nozzles with a uniform exit velocity profile. If suppresser nozzles are used to promote rapid mixing and reduce the dynamic pressure and temperature in the jet stream when it impinges on the ground, the method of Ref. 2 [Eqs. (2.2) and (2.4)] should be used to estimate the out-of-ground-effect lift loss.

A. Effect of Wing Height

Equation (2.2) applies to flat-plate configurations with the jet nozzle exit in the same plane as the lower surface. Reference 2 showed that if the nozzle is a distance Δh below the lower surface the lift loss is reduced because the entrainment action of the jet is further from the surface. For high wing configurations (Fig. 2.5) the lift loss is estimated for both the body alone (b) and for the wing body configuration (wb), assuming they are planar configuration. The lift loss

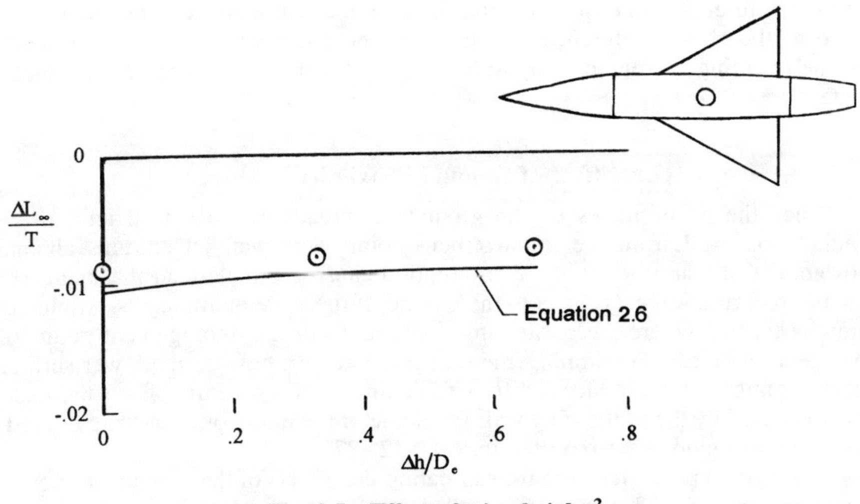

Fig. 2.5 Effect of wing height.[2]

for the high wing configuration (hw) is given by

$$\left(\frac{\Delta L_\infty}{T}\right)_{\text{hw}} = \left(\frac{\Delta L_\infty}{T}\right)_b + \left[\left(\frac{\Delta L_\infty}{T}\right)_{\text{wb}} - \left(\frac{\Delta L_\infty}{T}\right)_b\right]\left(1 - 0.4\sqrt{\frac{\Delta h}{D_e}}\right) \tag{2.6}$$

B. Pitching Moment

The pitching-moment induced out-of-ground effect is small and can often be ignored. If the front-to-rear area distribution is significantly asymmetrical, an approximation of the moment can be made by assuming the lift loss acts at the geometric center of the total planform area.

$$\frac{\Delta M_\infty}{TD_e} = \frac{\Delta L_\infty}{T}\frac{x_c}{D_e} \tag{2.7}$$

where ΔM_∞ is the increment of pitching-moment induced out-of-ground effect and x_c is the distance from the center of gravity to the geometric center of the planform area.

To achieve consistency with conventional pitching-moment coefficients, it can be argued that the pitching moments should be nondimensionalized using a representative wing chord. However, in hover and transition flight the direct jet lift and its jet-induced effects dominate the lift and pitching moment of the aircraft. In Ref. 1, the jet diameter was found to be appropriate to correlate hover data for a single jet configuration. For multiple jet configurations, it was found in Ref. 2 that if the areas of all of the jets are summed to form the total jet area A_j, then a characteristic length dimension was determined by considering A_j as a circle and then using its diameter D_e.

If a user of these methods wants a different reference length, the moment coefficients estimated by the methods presented here can be converted to another chosen reference dimension by multiplying the estimated coefficient by the ratio of the chosen reference dimension to the effective diameter D_e. Fidelity of the resulting correlation should be validated using appropriate experimental data.

II. Effect of Ground Proximity in Hover

When the jet impinges on the ground, it spreads out in a wall jet flowing radially outward from the impingement point. This wall jet entrains air and lowers the pressure below the configuration causing a download that increases as the configuration gets closer to the ground. If there are multiple jets, a fountain flow is formed where the outflowing wall jets from the impingement points of adjacent jets meet. The impingement of this fountain flow on the lower surface of the configuration produces a lifting force that at least partially offsets the suck-down caused by the outflowing wall jet. Some of the numerous hovering ground-effect investigations are reported in Refs. 12–27.

The method presented here for estimating the effects of the ground was developed in Ref. 15 and applies to vertical jets in steady-state hover conditions.

During a descent, the time taken for the overall flowfield to change, especially entrainment effects, causes the lift loss to lag the steady-state value; for example, the lift loss experienced at 18-ft height would be measured at about 20-ft steady state. Unfortunately a method for estimating these effects has not been published.

A. Single Jet Suckdown

The suckdown generated by a single jet impinging on the ground is caused by the entrainment action of the wall jet flowing radially outward from the impingement point as depicted in Fig. 2.6. The air that is entrained by the wall jet is drawn in through the gap between the ground and the edge of the configuration or blocking surface. As the height above the ground is reduced, the gap gets smaller, and the inflow velocity and suckdown pressures are increased. The suckdown pressures are highest toward the edge of the plate, or blocking surface (right side of Fig. 2.7), because the thickening of the outward flowing wall jet causes the passage for the entrained flow, between the plate and the wall jet, to expand as the entrained air flows inward. If the edge of the plate is sharp, the downward entrained airflow separates at the plate edge and produces a separation vortex that further increases suckdown.

As the plate is brought closer to the ground, a height can be reached at which the pressure distribution changes radically. Below this "critical height" (left side of Fig. 2.7), a "trapped vortex" is formed when the configuration, or blocking surface, is brought low enough that it almost reaches the upper edge of the outward flowing wall jet. As this condition is approached, the wall jet's "appetite" for entrained flow is satisfied by air pulled off of the outer regions of the upper surface of the wall jet itself, and the pressure distribution changes radically. This condition is seldom reached in practical configurations, but the possibility of encountering it can be checked by estimating the height of its onset.

The height, h_{tv}, below which the trapped vortex develops is given by:

$$\frac{h_{tv}}{d} \leq \frac{0.2(D-d)}{d} \tag{2.8}$$

where d is the jet diameter and D is the plate diameter or the equivalent diameter of the planform.

Pitching moments are particularly sensitive to the development of the trapped vortex under configurations that have fore and aft asymmetry.

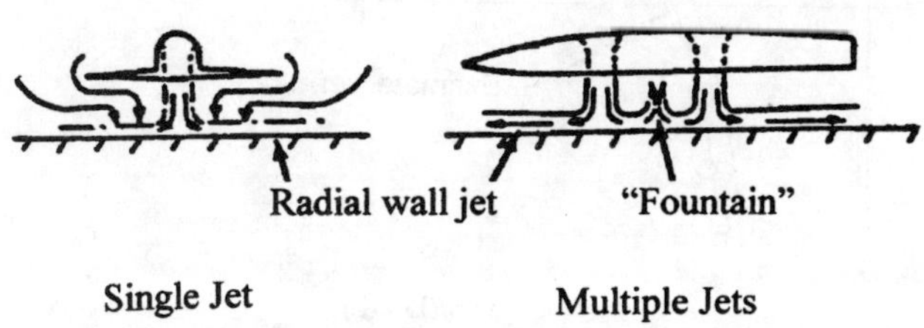

Fig. 2.6 Wall jet and fountain formation in-ground effect.

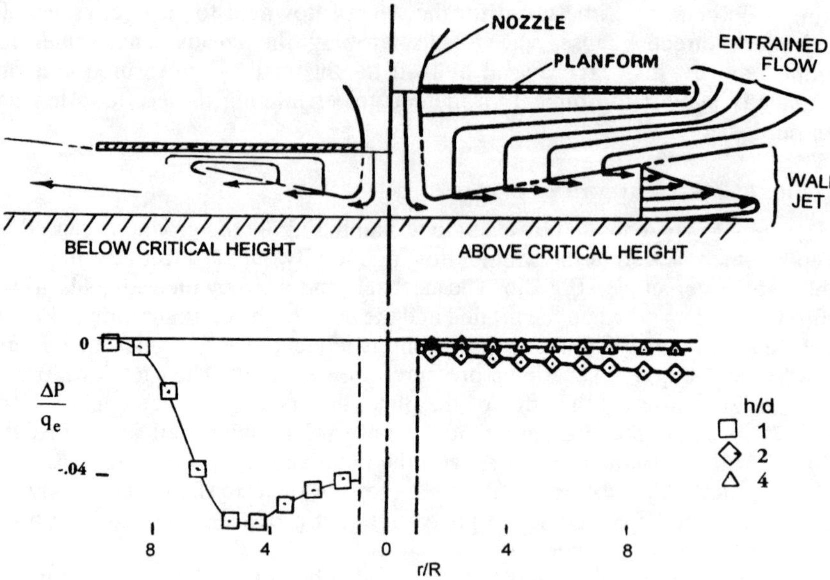

Fig. 2.7 Radial distribution of pressures induced on a circular plate.

Prior to the investigations of Refs. 10–12, the method of Ref. 13 was generally accepted as the method for estimating ground effects. Reference 13 demonstrated that the lift loss for different diameter plates could be correlated by using the height parameter $h/(D - d)$. Unfortunately, as reported in Refs. 10–12, the estimating method presented in Ref. 13 underestimates the lift loss (Fig. 2.8).

The present method is based on the extensive pressure distribution data of Ref. 14 and was developed and presented in Refs. 15 and 16. In developing

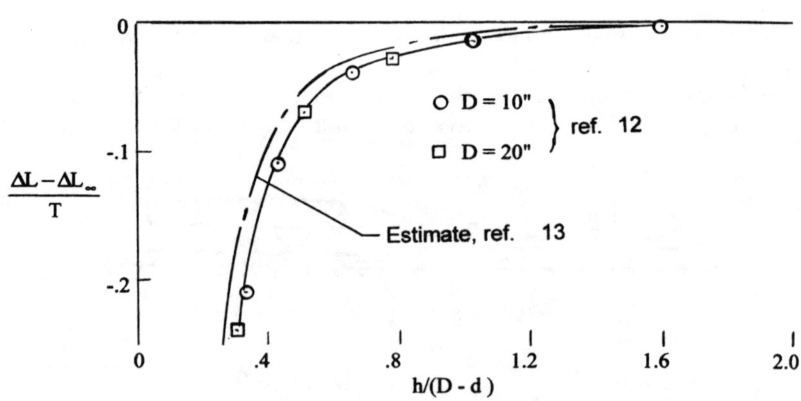

Fig. 2.8 Single jet lift loss in-ground effect.

the estimating method for a single jet configuration, the average pressure (defined as the lift divided by the area) induced by a centrally located jet in a circular plate is used. The pressure coefficient is defined as this average pressure divided by the jet dynamic pressure, defined as $q_e = T/2A_j$. Although this expression is appropriate only for subsonic jets, it has been used successfully for NPR as large as four. The method was developed using the average pressure data correlated from a range of configurations. The method presents the additional lift loss from ground proximity as

$$\frac{\Delta L - \Delta L_\infty}{T} = \left(C_{\text{p,ave}} - C_{p,\infty}\right)\frac{S}{2A_j} \tag{2.9}$$

where the pressure coefficient increment is given by

$$\left(C_{\text{p,ave}} - C_{p,\infty}\right) = K_{\text{sj}}\left(\frac{h}{D-d}\right)^{\text{exp}} \tag{2.10}$$

the constant K_{sj} is given by

$$K_{\text{sj}} = -0.043\frac{(\text{NPR})^{-0.1}(f_p)^{0.13}}{(S/A_j)} \tag{2.11}$$

and the exponent exp is given by

$$\text{exp} = -2.3(\text{NPR})^{-0.1}(f_p)^{0.14} \tag{2.12}$$

where f_p is the fineness ratio of planform.

If the planform does not have fore and aft symmetry, a pitching moment will be experienced in-ground effect. This pitching moment is estimated by considering the contributions of the suckdown on the area ahead of the jet and the area aft of the jet separately. The lift increment estimated for each area is multiplied by the effective arm at which this lift increment acts, and the increments are summed to obtain the net pitching moment:

$$\frac{\Delta M}{Td} = \frac{\Delta L_{\text{fwd}}}{T}\frac{x_{\text{e,fwd}}}{d} + \frac{\Delta L_{\text{aft}}}{T}\frac{x_{\text{e,aft}}}{d} \tag{2.13}$$

The effective arm depends on the height above the ground as illustrated in Fig. 2.9. Above the height equal to the distance x_s from the moment reference point to the center of area under consideration (the area forward or aft of the jet), the effective arm is taken as the geometric arm to the center of area:

$$x_e = x_s \tag{2.14}$$

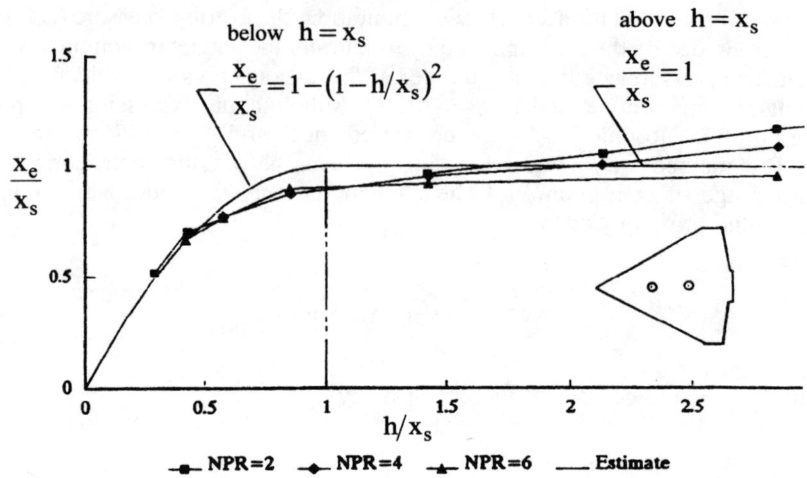

Fig. 2.9 Effect of height on the effective moment arm for single jet configuration moment estimates.

Below the height x_s, the effective arm decreases with height and is given by

$$x_e = x_s\left[1 - (1 - h/x_s)^2\right] \tag{2.15}$$

As shown in Fig. 2.10, the method does not apply below the height at which the trapped vortex [Eq. (2.8)] condition is encountered.

B. Multiple Jets and Fountain Effects

The suckdown and fountain effects generated by multiple jet configurations in ground effect are highly configuration dependent. The method presented here for estimating these effects is based on jet-induced pressure distributions taken on the extensive family of two, three, and four jet configurations studied in Refs. 14–16. In these studies the three and four jet configurations were rectangular arrangements generated by replacing the fore or aft jets of tandem pairs with side-by-side pairs.

When the wall jets, flowing radially outward from the impingement points of adjacent jets, meet, an upflow, or fountain, is formed (Fig. 2.11). The impingement of this fountain flow on the lower surface of the configuration tends to offset the wall-jet-induced suckdown.

Unfortunately the flowfield that includes this fountain flow also generates vortex-like flows between the upward flowing fountain and the downward-flowing jets (Fig. 2.12). These vortex-like flows induce high suction pressures, between the jets and the fountain, higher than would be induced by the wall jets alone. This additional suckdown can, in some cases, exceed the lift of the fountain.

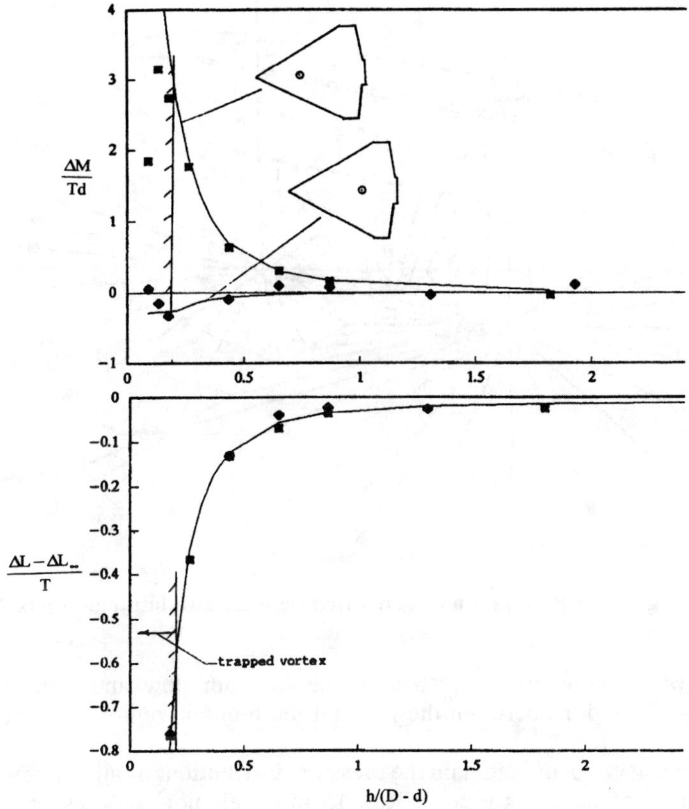

Fig. 2.10 Comparison of estimates with data for two jet locations.[15]

1. Basis of Method

Early attempts to develop methods for estimating the induced effects on multiple jet configurations started from an estimate for an equivalent single jet (i.e., Ref. 18). That is, the method estimated the suckdown that would be induced by a single jet having the same area as the total of all jets and makes adjustments to account for the number and distribution of the multiple jets relative to each other and to the planform. This approach tacitly assumes that the pressures induced outboard of the jet arrangement would be similar to those induced by a single large jet. The data and analysis of Refs. 14–16 have shown that this is not the case. The suckdown pressures outboard of the multiple jets are generally more negative and have a different slope with radial station than on single jet configurations. Apparently the vortex action between the fountain and the jets, shown in Fig. 2.12, also affects the pressures outboard of the jets.

The approach taken in the present method is illustrated, for a two jet configuration, in Fig. 2.13. The configuration is broken down into five regions: the

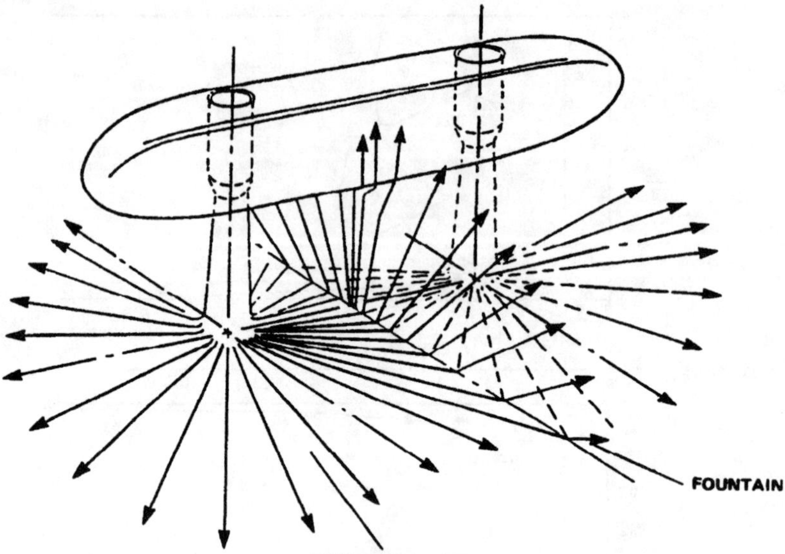

Fig. 2.11 Fountain flow generated between two impinging jets.

positive pressure region generated by the fountain flow and four suckdown regions—two regions between the jets and the fountain region and two regions outboard of the jets.

The typical effect of height on the pressure distributions along the centerline of one of the configurations used in Ref. 15 in developing the present method is shown in Fig. 2.14. The width of the fountain region decreases, and the fountain and suckdown pressures increase as the ground is approached.

Contour plots of the pressure distribution taken on another of the configurations of Ref. 14, shown in Fig. 2.15, demonstrate that the pressures are far from constant in the spanwise direction.

In developing the method presented here, the pressures in each region were integrated to determine the lift and pitching-moment contribution of each region for a wide range of configurations. The lift increments for each region are expressed in terms of the average pressure coefficient (based on the jet dynamic pressure) over the region, multiplied by the area of the region and the jet dynamic pressure:

$$\Delta L = \Delta S \frac{T}{2A_j} C_{\mathrm{p,ave}} \tag{2.16}$$

The net lift is taken as the sum of the lift increments:

$$\frac{\Delta L}{T} = \frac{\Delta L_{s,1}}{T} + \frac{\Delta L_{s,2}}{T} + \frac{\Delta L_f}{T} + \frac{\Delta L_{s,3}}{T} + \frac{\Delta L_{s,4}}{T} \tag{2.17}$$

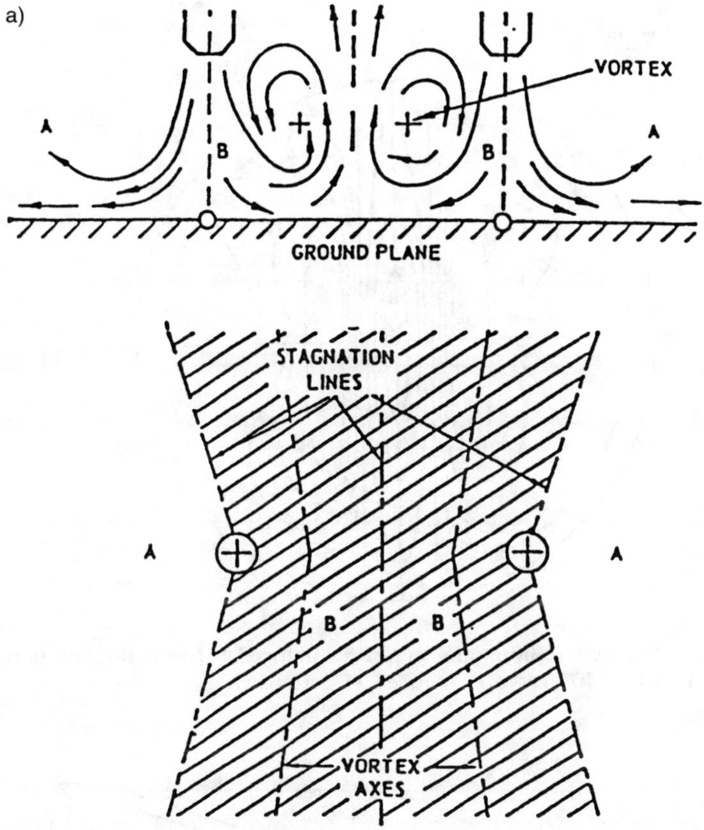

Fig. 2.12a Vortex-like flows and pressures induced between the fountain and the jets.[17] a) Flowfield b) Pressures induced.

and the pitching moment is the sum of the corresponding moment increments, which are each defined as the lift increment multiplied by the effective arm for that region.

The average pressure coefficient was determined for each region and plotted as a function of height. Figure 2.16 shows a typical variation of fountain pressure coefficient with height (in this case for the fountain region).

The break in the variation with height indicates the "top" of the fountain. The fountain pressure does not go immediately to zero above this top because, as flow visualizations have shown, the fountain is very unsteady in both position and heights (Fig. 2.17). At heights above the break in the curve, we are seeing the effects of the unsteady top of the fountain.

Studies of the fountain flowfield (Refs. 18 and 19) have shown that the fountain decay and spreading rates are greater than those observed for freejets or for wall jets. These observations are based on time-averaged surveys of the flow in

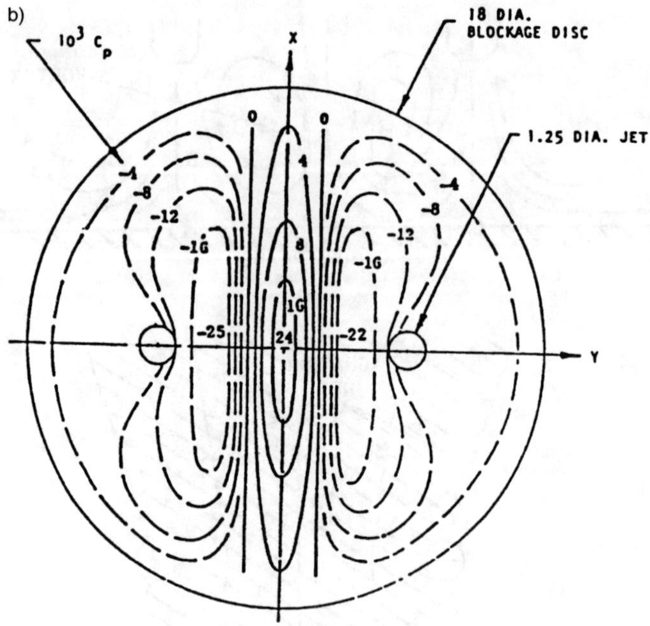

Fig. 2.12b Vortex-like flows and pressures induced between the fountain and the jets.[17] a) Flowfield b) Pressures induced. (Continued)

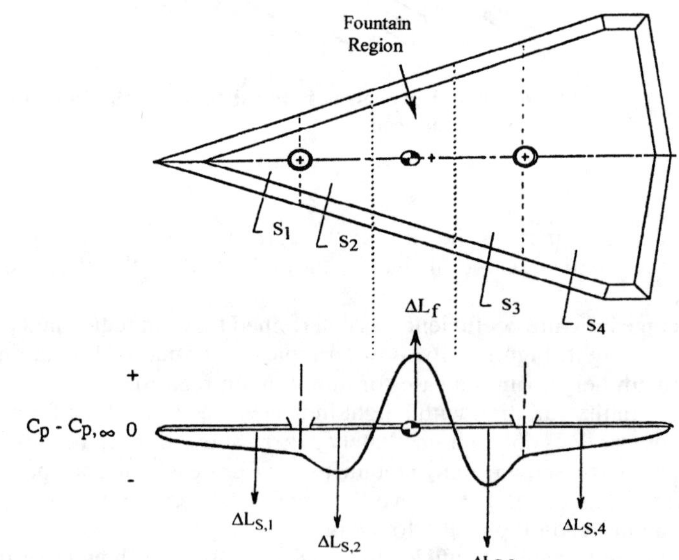

Fig. 2.13 Illustration of the method showing the fountain and suckdown regions.

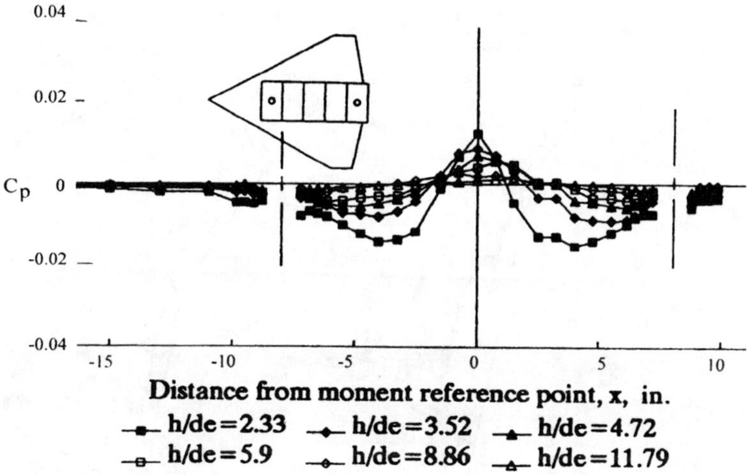

Fig. 2.14 Effect of height on the centerline pressure distribution for one of the configurations of Ref. 15.

the fountain region. Time-history measurements of the pressures in the fountain impingement region show that these pressures are very unsteady,[20] and the time-history photographs of Ref. 21 show that the unsteady fountain affects the entire flowfield between the jets as well as the mixing region of the jets themselves (Fig. 2.18).

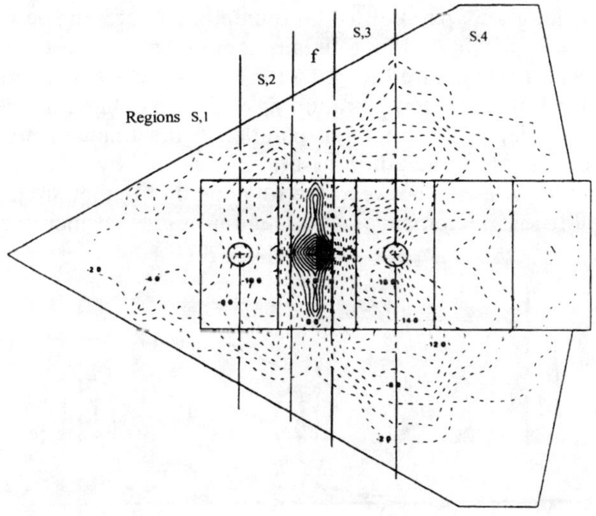

Fig. 2.15 Typical pressure contour plot showing positive (——) and negative (- - - -) pressures (from Fig. 39 of Ref. 14).

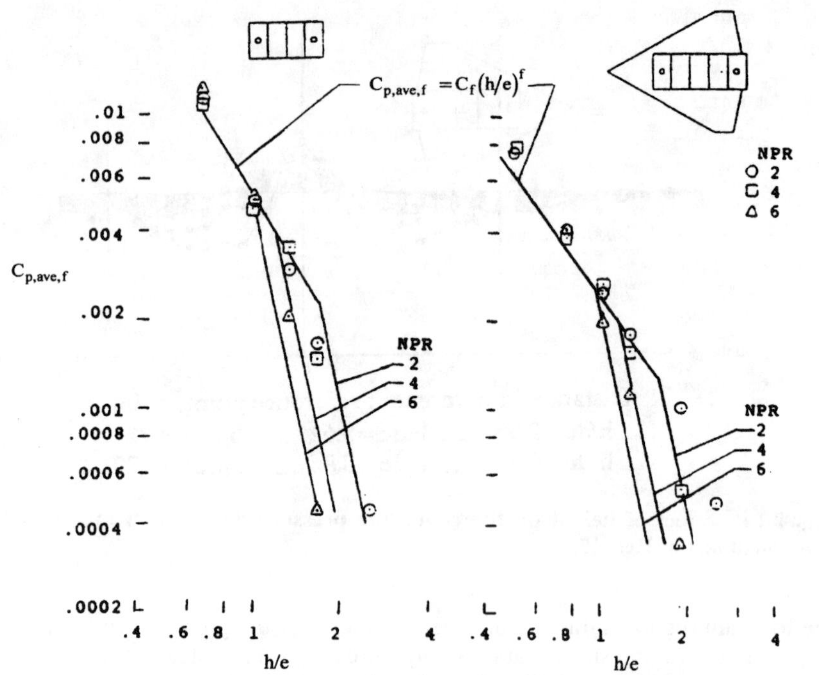

Fig. 2.16 Typical variation of average fountain pressure coefficient with height.

a. Effect of noncircular jets. When circular jets are replaced by rectangular jets, with their long axis parallel to the fountain, Ref. 15 showed that both the pressures and the width of the fountain region are increased at the lowest heights. The width is increased by a factor $K_{n,x}$, which is a function of the jet spacing e and the nozzle fineness ration f_n, which is defined as the ratio of the length of the nozzle, in the direction parallel to the fountain, to the width of the nozzle. Similarly the fountain pressure is increased by a factor $K_{n,f}$.

The suckdown pressures between the fountain and the jet are also increased; however, the lift loss increment for this region might not increase because the

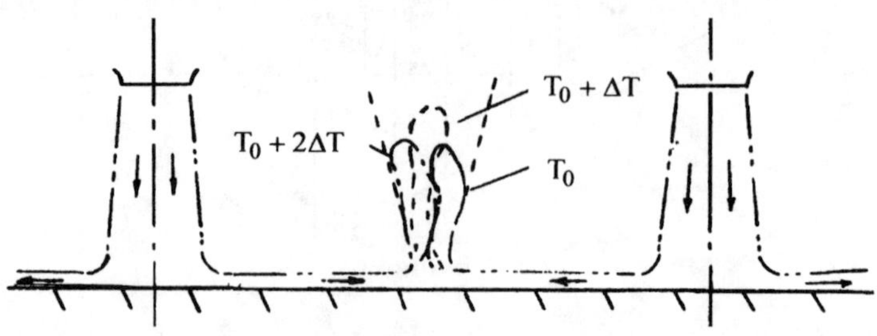

Fig. 2.17 Schematic of the unsteady fountain flow.

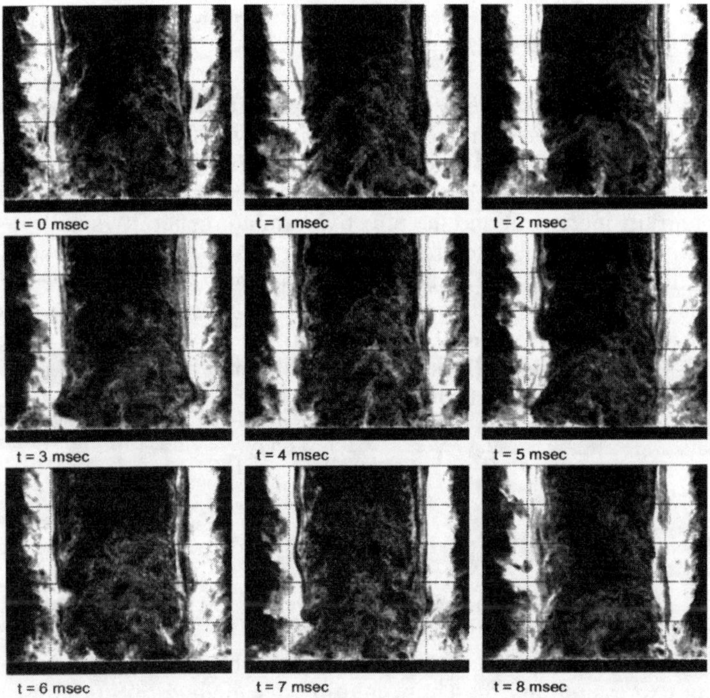

Fig. 2.18 High-speed photographs of the unsteady fountain flow between two vertically impinging jets: $s/d = 3, h/d = 7$ (Ref. 21).

increased width of the fountain region reduces the area of this suckdown region. The factor $K_{n,s,inb}$ by which the pressures are increased is a function of the nozzle fineness ratio and the half-width Y at the nozzle. Outboard of the jets the pressures are reduced by the factor $K_{n,s,out}$.

The net result is that noncircular jets properly oriented with respect to the configuration can reduce the suckdown in-ground effect.

b. Three and four jet configurations. The net suckdown for three and four jet configurations is less than that for the corresponding two jet configuration as would be expected; however, this reduction is not caused by an increase in fountain lift. The fountain contribution is actually reduced at all but the lowest heights. The reduction in net suckdown is caused by a reduction in the suckdown generated between the fountain and the jet. This reduction in suckdown again emphasizes the importance of the vortex flows generated between the fountain and the jet. The lift losses outboard of the three jet configurations are similar to those outboard of a corresponding two jet configuration.

Estimates for three and four jet configurations are made by breaking the configuration down into the same regions shown in Fig. 2.13. Additional terms, which include the effect of the length/width ratio E, of the jet pattern, are introduced to account for the effects of the side-by-side pair of jets replacing the individual jet in the fore and/or aft jet positions.

c. Effect of differential thrust. The effect of increasing the rear thrust relative to the front thrust is to move the fountain forward, as expected (Fig. 2.19), and the amount of movement increases with height. However the effect on the fountain contribution to pitching moment is generally negligible because both the fountain lift increment and the effective arm are generally small.

The primary effect of differential thrust is on the suckdown increments, particularly those between the fountain and the jets. When differential thrust moves the fountain region forward, the size, and therefore the lift-loss contribution of the forward region is reduced, and the lift loss of the aft region is increased. The change in lift loss, for small differentials, is directly proportional to the ratio of the local thrust to the average thrust and can be estimated by:

$$\left(\frac{\Delta L}{T}\right)_{dT} = \left(\frac{T_l}{T/2}\right)\left(\frac{\Delta L}{T}\right)_{1.0} \tag{2.18}$$

where T_l is the thrust of the local jet (or side-by-side pair) nearest the suckdown region.

The available data do not indicate the level of local thrust below which the preceding expression does not apply. If, for example, for a two jet configuration the front jet thrust goes to zero, the problem becomes a single jet case, and the preceding expression and the entire multiple jet method do not apply. Instead the single jet method discussed earlier must be used.

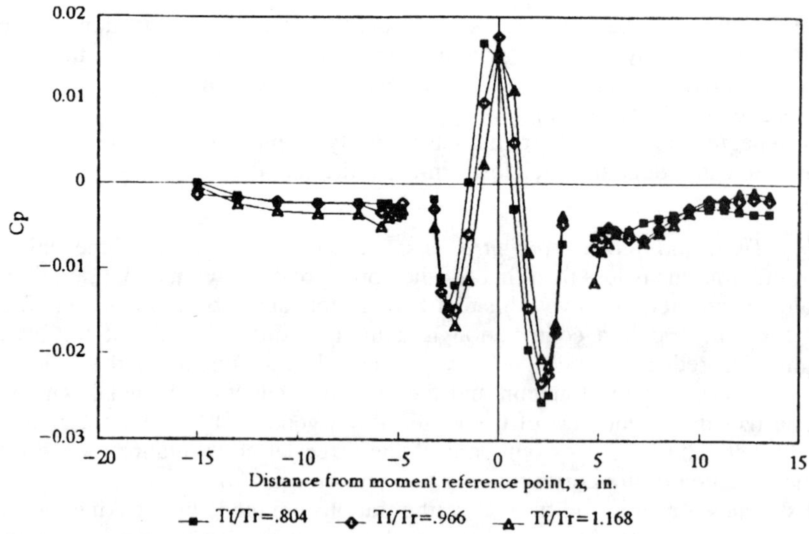

Fig. 2.19 Effect of differential thrust on centerline pressure distribution.

d. Effect of body contour. If the lower surface of the body is not flat, the contribution of the fountain flow will be reduced because, as indicated in the sketch at the top of Fig. 2.20, not all of the fountain flow will be stopped and turned to the horizontal.

Unfortunately little systematic data are available on the effects of body shape. The analysis of available data (Ref. 23) shows a major effect of curvature (or corner radius) and of the orientation of the fountain flow with respect to the body. As shown in Fig. 2.20, a two jet configuration with a lengthwise fountain (parallel to the body) experiences a much larger reduction in fountain lift than a configuration is which the primary fountain is crosswise.

For two jet configurations the curvature is assumed to affect only the fountain lift, and the fountain contribution is given by

$$\left(\frac{\Delta L_f}{T}\right) = K_r\left(\frac{\Delta L_f}{T}\right)_{r=0} \tag{2.19}$$

For configurations with three or four jets, the fountain lift is reduced as just shown, and, in addition, the suckdown between the fountain and the side-by-side pair of jets is also affected. Specifically the reduction in suckdown caused

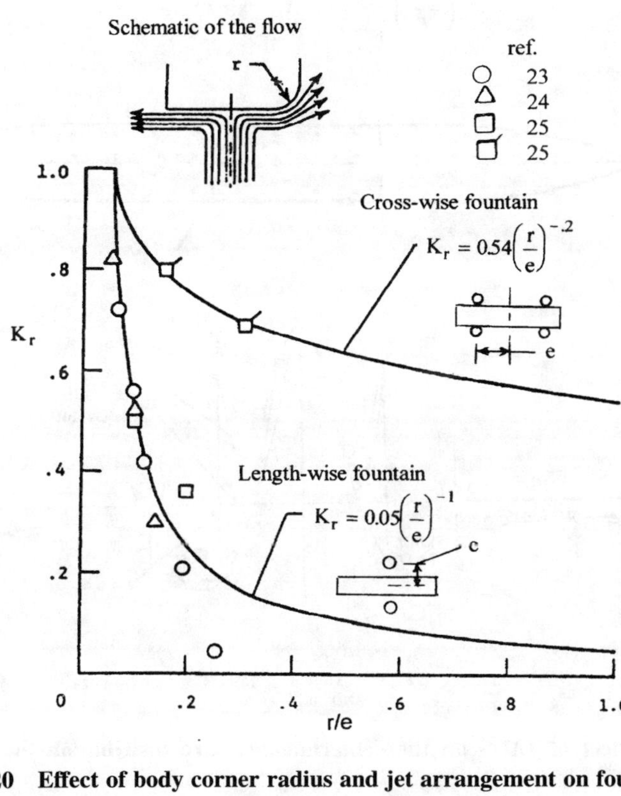

Fig. 2.20 Effect of body corner radius and jet arrangement on fountain lift.

by the nearby pair is reduced by the curvature and estimated by multiplying by the curvature factor K_r.

e. Effect of LIDs. Frequently strakes, or lift improvement devices (LIDs) are installed on the lower surface in an attempt to increase the fountain lift (or minimize hot-gas ingestion). The data of Ref. 26 show that fountain lifting pressures are increased only at the lowest heights. The primary effect of the LIDs (Fig. 2.21) is not to increase the positive pressures in the fountain impingement region but to reduce the suckdown pressures induced between the fountain and the jets. As shown in Fig. 2.21, the high suction pressures induced here are greatly reduced and at some points changed to positive pressures.

f. Effect of jet deflection. The data of Ref. 27 showed that, for single jet configurations, the suckdown is reduced by deflecting the jet away from the vertical. Similar data are not available on the effects of deflection on the fountain contribution. However because, on most configurations, the suckdown tends to dominate, the present method assumes that the expression developed in Ref. 27 applies (Fig. 2.22):

$$\left(\frac{\Delta L}{T}\right)_\delta = \left(\frac{\Delta L}{T}\right)_{\delta=0} \sin^2\delta \tag{2.20}$$

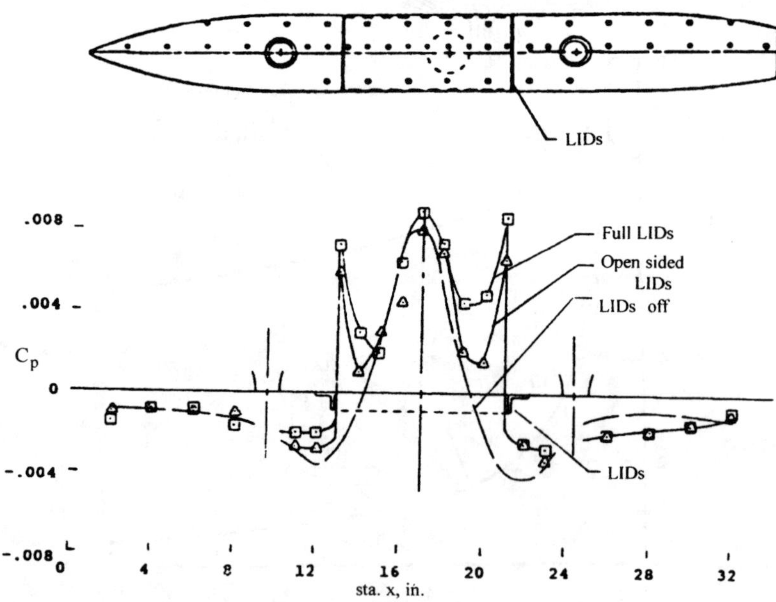

Fig. 2.21 Effect of LIDS on the centerline pressure distribution for a two jet configuration.[26]

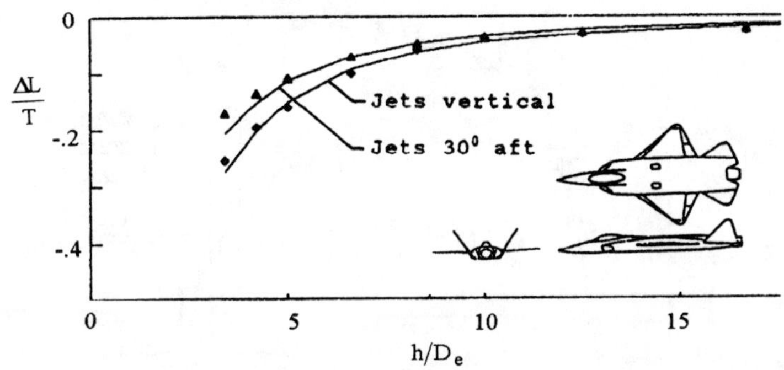

Fig. 2.22 Effect of jet deflection on the net suckdown for the two jet configuration of Ref. 27.

g. Effect of wing height. The fountain and fountain-related effects have been found to be felt only on the body of a high wing configuration. The wing contribution can be accounted for by assuming that the wing panels outboard of the body experience only the suckdown that would be produced by an equivalent single jet. The method calculates the net wing suckdown at $(h + \Delta h)$ by subtracting the estimated suckdown for the wing center section (that part of the wing that overlaps the body) from the estimate for the full wing. This is added to the body contribution (which includes the fountain-related effects) to obtain the total jet induced lift on the configuration (Fig. 2.23).

h. Effect of edge shape. The shape of the edge of the plate used to represent the planform was one of the factors investigated in Refs. 29 and 30 in attempting

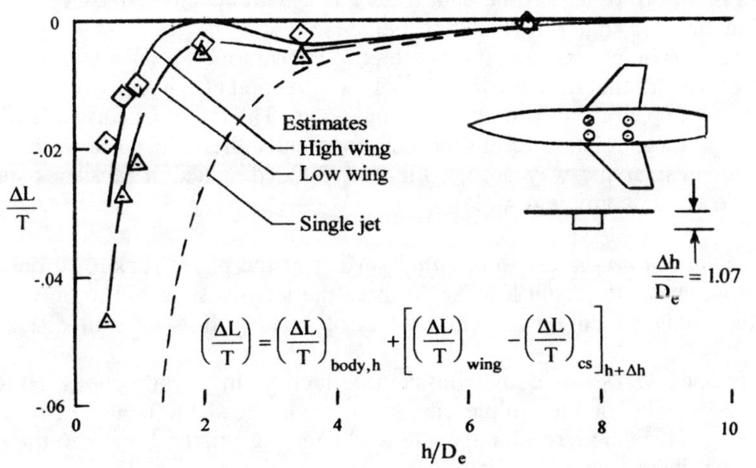

$$\left(\frac{\Delta L}{T}\right) = \left(\frac{\Delta L}{T}\right)_{body,h} + \left[\left(\frac{\Delta L}{T}\right)_{wing} - \left(\frac{\Delta L}{T}\right)_{cs}\right]_{h+\Delta h}$$

Fig. 2.23 Effect of wing height on the net suckdown for the four jet configuration of Ref. 28.

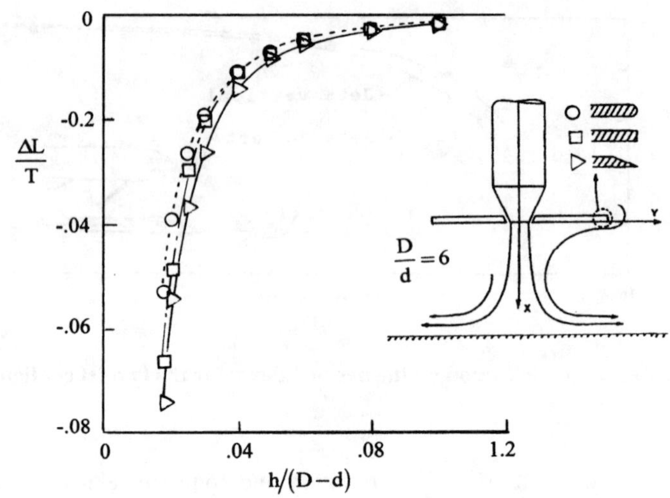

Fig. 2.24 Effect of edge shape on suckdown (Ref. 29).

to find the reason for the difference between the large amount of recent suckdown data and that estimated by the previously accepted method of Ref. 13 (see Fig. 2.8). Both investigations found that a round edge (as used in Ref. 13) significantly reduced the suckdown (Fig. 2.24) relative to that measured for a plate with a beveled edge.

The radius used on the round-edged plates was large, about 10% of the jet diameter, and an order of magnitude larger than the leading-edge radius of the airfoil of a wing likely to be used on a fighter aircraft. The method presented here is therefore based on the data from a test with beveled edges, and a factor for adjusting for edge effects is not included.

It can also be observed that this effect of curvature correlates with the observance of Ref. 15 that the contributions of the fore- and afterbody parts of the fuselage do not appear to contribute to the suckdown. The radius of curvature of these sections is large relative to the jet, and apparently the induced downflow stays attached most of the way around the body, and the suction pressures induced on the lower side are negligible.

i. Effect of nozzle pressure ratio. Most of the hover suckdown data have been obtained with jets at low NPR where the jet was subsonic or only slightly into the supercritical range. The effects of higher NPR were investigated in Refs. 29–31.

Reference 29 showed a significant reduction in suckdown as NPR was increased to six for the simple case of a single jet at the center of a circular plate (Fig. 2.25). The results also showed some surprising breaks in the curve, which schlieren studies showed to be associated with the impingement on the ground altering the shock diamond pattern in the jet stream at certain heights.

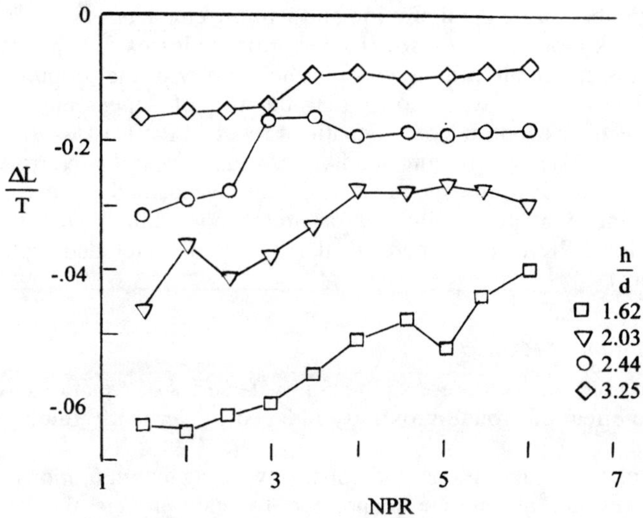

Fig. 2.25 Effect of supercritical NPR on suckdown induced by a single jet on a circular plate.[29]

The effects of higher NPR on a twin jet configuration with two rectangular jets[27] produce only a small decrease in suckdown as a result of higher NPR (Fig. 2.26).

The reasons for the difference between the single jet and the twin jet result are not known. Experiments by students of Shih et al.[31] have shown the presence of unsteady feedback between the nozzle exit and the ground plane, which is very

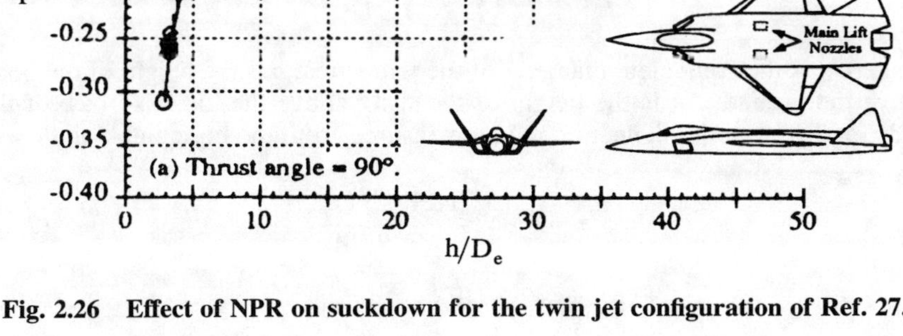

Fig. 2.26 Effect of NPR on suckdown for the twin jet configuration of Ref. 27.

dependent on the nozzle configuration and its flow characteristics. The presence of this feedback loop can cause several performance losses such as lift loss, high dynamic loads on the aircraft structure in the vicinity of the propulsion system, and ground erosion. It was shown that the use of supersonic microjets to disrupt the feedback loop in some situations reduced the lift loss, the near-field noise, and the unsteady pressure loads. Unfortunately the performance losses were not uniform over the range of variables tested, and further research is needed to achieve practical solutions for aircraft application. Therefore a factor to account for NPR in the supercritical range is not included in the method described next.

2. *Presentation of Method*

The following sections presents the complete series of equations available to estimate the effect of ground proximity induced by vertically impinging jets in hover.

For flat-plate configurations and those low wing configurations where the lower surfaces are all in one plane, the fountain affects the entire lower surface, and the section for the body increment is used. On high wing configurations the fountain affects only the body, and the wing increment is estimated as that produced by an equivalent single jet.

a. Wing increment. The net suckdown for a high wing configuration is given by

$$\left(\frac{\Delta L}{T}\right) = \left\{\left(\frac{\Delta L}{T}\right)_{b,h} + \left[\left(\frac{\Delta L}{T}\right)_{w} - \left(\frac{\Delta L}{T}\right)_{\text{cs}}\right]_{h+\Delta h}\right\} \sin^2\delta \tag{2.21}$$

where $(\Delta L/T)_b$ is the body contribution presented later, which includes the fountain effects and δ is the jet deflection angle measured from the horizontal.

The suckdown that would be experienced by the full wing is given by

$$\left(\frac{\Delta L}{T}\right)_w = C_{p,w}\frac{S_w}{2A_j} \tag{2.22}$$

where S_w is the wing area, and the pressure coefficient is given by

$$C_{p,w} = K_{\text{sj,w}}\left(\frac{h+\Delta h}{D_w - D_e}\right)^{\exp_w} \sin^2\delta \tag{2.23}$$

where D_w is the equivalent diameter of the wing area, h is the height of the body above the ground, Δh is the height of the wing above the lower surface of the body, and δ is the jet deflection angle, measured from the horizontal, and

$$K_{\text{sj,w}} = -0.043\frac{(\text{NPR}^{-0.1}(f_{p,w})^{0.13}}{S_w/A_j} \tag{2.24}$$

$$\exp_w = -2.3(\text{NPR})^{-0.1}(f_{p,w})^{0.14} \tag{2.25}$$

The suckdown associated with the center section is given by

$$\left(\frac{\Delta L}{T}\right)_{\text{cs}} = C_{\text{p,cs}} \frac{S_{\text{cs}}}{2A_j} \tag{2.26}$$

where the pressure coefficient is given by

$$C_{\text{p,cs}} = K_{\text{sj,w}} \left(\frac{h + \Delta h}{D_{\text{cs}} - D_e}\right)^{\exp_{\text{cs}}} \sin^2 \delta \tag{2.27}$$

and where S_{cs} is the area of the center section and D_{cs} is the equivalent diameter of the center section area (the center section is defined as that part of the wing that overlaps the body planform) and

$$K_{\text{sj,cs}} = -0.043 \frac{(\text{NPR})^{-0.1}(f_{\text{p,cs}})^{0.13}}{S_{\text{cs}}/A_j} \tag{2.28}$$

and

$$\exp_{\text{cs}} = -2.3(\text{NPR})^{-0.1}(f_{\text{p,cs}})^{0.14} \tag{2.29}$$

The moment contribution of the wing is given by

$$\left(\frac{\Delta M}{TD_e}\right) = \left(\frac{\Delta M}{TD_e}\right)_{b,h} + \frac{X_w}{D_e}\left[\left(\frac{\Delta L}{T}\right)_w - \left(\frac{\Delta L}{T}\right)_{\text{cs}}\right]_{h+\Delta h} \tag{2.30}$$

where X_w is the distance from the moment reference point to the center of area of the outer wing panels.

For wings with dihedral, the wing height is measured at the spanwise station of the wing mean geometric chord.

b. Body contribution. One of the primary problems that can be encountered in applying the method is that of how to define the body planform. The body swoops upward at the nose and tail, and the body cross sections, at least those fore and aft of the wing, tend toward being circular or elliptical is shape. Reference 15 has suggested that these do not significantly contribute to the suckdown and should be ignored in deciding on the planform area and fineness ratio to be used in estimating the body contributions. Judgement must be used. If the body has a substantially flat portion, with side corner radius that can be defined, these should be used in defining the body area and fineness ratios. If not, only that portion of the body area that lies between the jets should be used.

In the present method the chosen body planform (or the entire configuration if it is a low wing or flat-plate configuration) is broken into five regions as shown in Fig. 2.13. The lift and associated moment increments are estimated for each region and summed. The net lift loss is given by

$$\left(\frac{\Delta L}{T}\right)_{\text{body}} = \left(\frac{\Delta L_{S_1}}{T} + \frac{\Delta L_{S_2}}{T} + \frac{\Delta L_f}{T} + \frac{\Delta L_{S_3}}{T} + \frac{\Delta L_{S_4}}{T}\right) + \left(\frac{\Delta L_{\text{LID}}}{T}\right) \tag{2.31}$$

where $(\Delta L_{\mathrm{LID}}/T)$ is the contribution of lift improvement devices, which are discussed later.

c. Fountain lift. The fountain lift increment is the average positive pressure induced by the fountain multiplied by the area over which it acts:

$$\frac{\Delta L_f}{T} = C_{p,f}\frac{S_f}{2A_j}K_{n,f}K_r \tag{2.32}$$

where $C_{\mathrm{p,ave,f}}$ is the average pressure coefficient presented later [Eq. (2.42)], S_f is the area over which this average pressure acts, and $K_{n,f}$ and K_r are the nozzle shape factor and the body shape factor, respectively.

The factor that accounts for noncircular nozzles is

$$K_{n,f} = f_n^{0.25}\left(\frac{h}{e}\right)^{0.004(f_n-1)(Y/d)^2} \tag{2.33}$$

and the body shape factor (see Fig. 2.20) is given by, for lengthwise fountains,

$$K_r = 0.05\left(\frac{r}{e}\right)^{-1} \tag{2.34}$$

and for spanwise fountains,

$$K_r = 0.54\left(\frac{r}{e}\right)^{-0.2} \tag{2.35}$$

The area S_f is the product of the average body width between the jets multiplied by the length $2X_0$, (in the x direction) of the area of positive pressures. That is,

$$S_f = 4X_0Y \tag{2.36}$$

where Y is the average half-body width between the jets. For configurations where the surface between the jets does not extend from jet to jet. (As in jets on either side of the body), the convention shown in Fig. 2.27 is adopted.

Expressions for X_0 were determined empirically from the data of Ref. 14. If the jets are close together ($e/d \leq 1.5$), the fountain positive pressure region extends from jet to jet, and

$$\frac{X_0}{e} = 0.36\left(\frac{Y_f+Y_r}{2d}\right)^{-0.08}\left(\frac{h}{e}\right)^{b}K_{n,x} \tag{2.37}$$

For jet spacing ratios less than $e/d = 1.5$, and at the lower operating heights, the half-width of the fountain region is given by

$$\frac{X_0}{e} = 1 \tag{2.38}$$

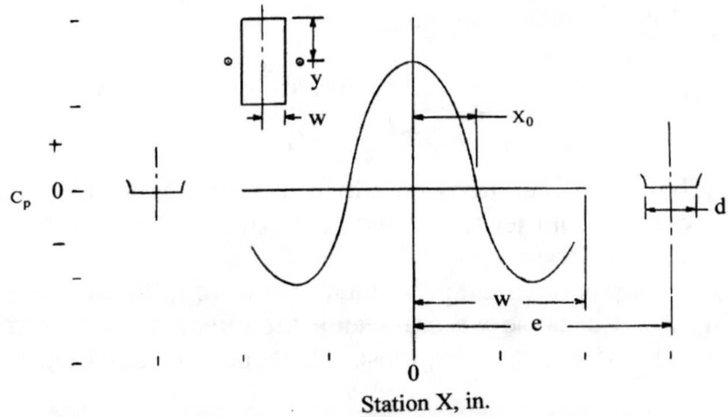

Fig. 2.27 Definition of terms for jets outside the body.

where Y_f is the body half-width at the front jet and Y_r is the body half-width at the rear jet and the constant $K_{n,x}$, which accounts for the effects of noncircular nozzles, and is given by

$$K_{n,x} = f_n^{0.25}\left(\frac{h}{D_e}\right)^{-0.08(f_n-1)} \tag{2.39}$$

and the exponent b is given by

$$b = 0.6\left(\frac{e}{d}\right)^{-0.16}\left(\frac{Y_f + Y_r}{2d}\right)^{0.25} \tag{2.40}$$

At the higher heights X_0 appears to reach a limit, which is taken as

$$\frac{X_0}{e} = 0.5 \tag{2.41}$$

In making the estimates of the fountain lift, the fountain half-width X_0 is taken as the lesser of that calculated from Eqs. (2.37) and (2.41).

The average fountain pressure is calculated in two height ranges. At the lower heights the average pressure coefficient is given by

$$C_{\text{p,ave,f}} = 0.16\left(\frac{S}{A_j}\right)^{-0.72}\left(\frac{e}{d}\right)^{-0.5}\left(\frac{Y_f + Y_r}{2d}\right)^{0.25}\left(\frac{w}{e}\right)^{0.5}\left(\frac{h}{e}\right)^{f} \tag{2.42}$$

where for values of $e/d \leq 3.3$ the exponent f is given by

$$f = -2.2\left(\frac{w}{e}\right)^{0.5} \tag{2.43}$$

and for values of e/d greater than 3.3

$$f = -4\left(\frac{e}{d}\right)^{-0.5}\left(\frac{w}{e}\right)^{0.5} \tag{2.44}$$

Also, for wing body configurations on which the wing chord at the side of the body does not extend from jet to jet, the value of w is taken as half the wing chord at the side of the body.

At the higher heights the average fountain pressure falls off at a much more rapid rate (Fig. 2.16) because we are seeing the effects of the unsteady "top" of the fountain. The height at which these effects become apparent is given by

$$\frac{h_f}{e} = 3.7(\mathrm{NPR})^{-0.5}\left(\frac{e}{d}\right)^{-0.2}\left(\frac{w}{e}\right)^{0.6(\mathrm{NPR})^{-0.6}} \tag{2.45}$$

Above h_f the average fountain pressure is given by

$$C_{\mathrm{p,ave,f}} = 0.16\left(\frac{S}{A_j}\right)^{-0.72}\left(\frac{e}{d}\right)^{-0.5}\left(\frac{Y_f + Y_r}{2d}\right)^{0.25}\left(\frac{w}{e}\right)^{0.5}\left(\frac{h}{e}\right)^{\mathrm{f}}\left(\frac{h_f}{h}\right)^{3} \tag{2.46}$$

d. Suckdown—two jet configurations. The general expression for the suckdown increment for each of the four suckdown regions (Fig. 2.13) is

$$\frac{\Delta L_s}{T} = C_{\mathrm{p,ave,s}}\frac{S_x}{2A_j}\frac{T_l}{T/2} \tag{2.47}$$

where T_l is the thrust of the local jet (or side-by-side pair of jets) nearest the region for which the suckdown is being estimated, T is the total thrust, and S_x is the area of the region for which the suckdown is being estimated.

For the region forward of the front jet and aft of the rear jet, S_x is the geometric area. For the regions between the jets and the fountain, S_x is the difference between the geometric area and half the fountain area S_f.

The average pressure coefficient $C_{\mathrm{p,ave,s}}$ is different for each suckdown region. Four sets of equations are required: one for the inboard regions at low heights, a second for the inboard regions at higher heights, and a third and fourth for the outboard regions at low and high heights. The average pressure coefficient is given by the following:

Inboard at low heights:

$$C_{\mathrm{p,ave,s}} = K_{\mathrm{n,s,inb}}K_{\mathrm{s,inb}}H_s^{g} \tag{2.48}$$

Inboard at the higher heights:

$$C_{\mathrm{p,ave,s}} = K_{\mathrm{n,s,inb}}K_{\mathrm{s,high}}H_s^{-1.8} \tag{2.49}$$

Outboard at low heights:

$$C_{\mathrm{p,ave,s}} = K_{\mathrm{n,s,out}} K_{\mathrm{s,out}} H_s^i \tag{2.50}$$

Outboard at the higher heights:

$$C_{\mathrm{p,ave,s}} = K_{\mathrm{n,s,out}} K_{\mathrm{s,high}} H_s^{-1.8} \tag{2.51}$$

where the height parameter H_s is given by

$$H_s = \frac{h}{D_p - D_e} \mathrm{NPR}^{\left(\frac{0.8}{Y/d}\right)} \tag{2.52}$$

and D_p is the effective diameter of the planform under consideration and Y is the half-width of the planform at the fountain. For low-wing configurations in which the chord of the wing, at the edge of the body, is less than the jet spacing, Y is taken as the average half-width between the front and rear jets in calculating H_s:

The constants used in Eqs. (2.48–2.51) are given by

$$K_{\mathrm{s,inb}} = -0.3\left(\frac{S}{A_j}\right)^{-1}\left(\frac{e}{d}\right)^{-0.15} \tag{2.53}$$

$$K_{\mathrm{s,\,high}} = -0.135\left(\frac{S}{A_{\mathrm{j}}}\right)^{-1}\left(\frac{e}{d}\right)^{0.5}\left(\frac{Y}{d}\right)^{-0.36}\left(\frac{w}{e}\right)^{1.34} \tag{2.54}$$

$$K_{\mathrm{s,out}} = -0.062\left(\frac{S}{A_j}\right)^{-0.84}\left(\frac{Y}{d}\right)^{0.25}\left(\frac{X_{\mathrm{s,out}}}{d}\right)^{-0.5} \tag{2.55}$$

$$K_{\mathrm{n,s,inb}} = f_n^{0.65}\left(\frac{Y}{d}\right)^{-0.25(f_n-1)}\left(\frac{h}{D_e}\right)^{0.0015(f_n-1)(Y/d)^2} \tag{2.56}$$

$$K_{\mathrm{n,s,out}} = f_n^{-0.5}\left(\frac{h}{D_e}\right)^{0.12(f_n-1)} \tag{2.57}$$

and the exponents in Eqs. (2.48) and (2.50) are given by

$$i = -0.96\left(\frac{e}{d}\right)^{-0.25}\left(\frac{X_{\mathrm{s,out}}}{d}\right)^{0.38} \tag{2.58}$$

$$g = -0.38\left(\frac{S}{A_j}\right)^{0.34}\left(\frac{e}{d}\right)^{-0.25}\left(\frac{Y}{d}\right)^{-0.15} \tag{2.59}$$

where Y is the half-width at the jet and $X_{\mathrm{s,out}}$ is the distance from the jet to the center of the area outboard of the jet.

e. Fountain—three and four jets. The overall net suckdown for three and four jet configurations is smaller than for two jet configurations. However the

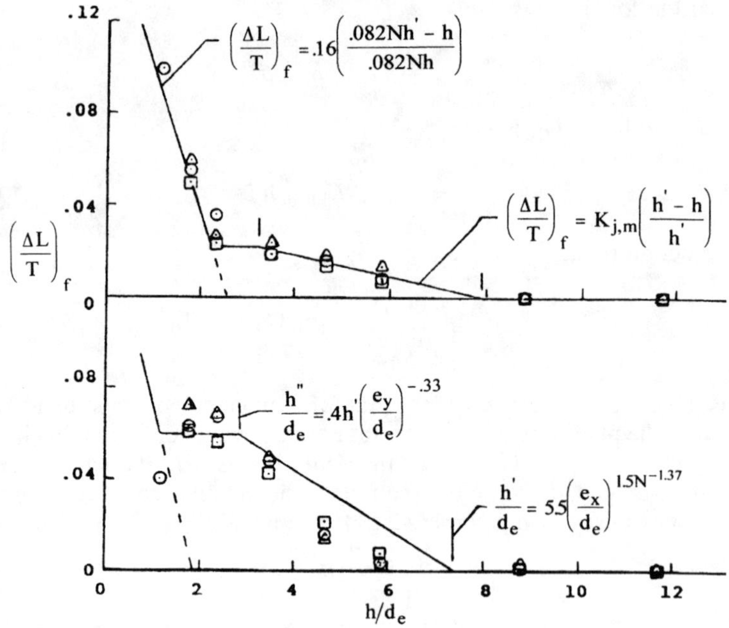

Fig. 2.28 Typical fountain lift increments induced on three and four jet configurations.[15]

reduction is not caused by an increase in fountain lift but to a reduction in suck-down. The fountain lift is actually reduced for many three and four jet configurations.

As expected, the fountain lift increment reduces with height; however, the rate is irregular, and the estimates are made in three regions (Fig. 2.28). In the first region, closest to the ground, the lift drops off very rapidly. This is followed by a region where the lift is relatively constant above which the lift drops off gradually to zero.

The height at which the fountain lift goes to zero is designated as h' and is given by

$$\frac{h'}{D_e} = 5.6\left(\frac{e_{\text{ave}}}{D_e}\right)^{0.85} E^{-0.3} \tag{2.60}$$

where the average jet spacing is given by

$$e_{\text{ave}} = \frac{2e_x + e_{y,f} + e_{y,r}}{N} \tag{2.61}$$

and the length/width ratio of the jet pattern is given by

$$E = \frac{e_x}{e_y} \tag{2.62}$$

where N is the number of jets and the value of e_y is the half-width at the front jet or the aft jet, whichever is greater.

Below h' (in the top region) the fountain lift is given by

$$\frac{\Delta L_f}{T} = 0.32\left(\frac{e_{\text{ave}}}{D_e}\right)^{-1}\left(\frac{S}{A_j}\right)^{-0.2}\left(\frac{h'-h}{h'}\right)K_r \tag{2.63}$$

This expression holds down to a height h'' given by

$$\frac{h''}{D_e} = 0.4\left(\frac{h'}{D_e}\right)\left(\frac{e_y}{D_e}\right)^{-0.33} \tag{2.64}$$

Below h'' (in the midregion) the fountain lift is given by

$$\frac{\Delta L_f}{T} = 0.32\left(\frac{e_{\text{ave}}}{D_e}\right)^{-1}\left(\frac{S}{A_j}\right)^{-0.2}\left(\frac{h'-h''}{h'}\right)K_r \tag{2.65}$$

At the very lowest heights the fountain lift increases rapidly as the ground is approached. In this region the fountain lift increment is given by

$$\frac{\Delta L_f}{T} = 0.16\left(\frac{0.082Nh'-h}{0.082Nh'}\right)K_r \tag{2.66}$$

Below a height of h'', the fountain lift is taken as the greater of that given by Eqs. (2.65) and (2.66).

f. Suckdown—three and four jets. For most three and four jet configurations the largest reduction in the net suckdown is the decrease in the suckdown induced between the primary fountain and the pair (or pairs) of side-by-side jets. The suckdown for three and four jet configurations is estimated by modifying the estimate for two jet configurations. The suckdown between the fountain and the side-by-side pair (in suckdown regions 2 or 3 of Fig. 2.13) is given by:

$$\frac{\Delta L_{s,2(\text{or}3)}}{T} = \left(\frac{\Delta L_s}{T}\right)_{2\text{-jet-est}} + \frac{\Delta\Delta L_{s,2(\text{or}3)}}{T} \tag{2.67}$$

where the subscript s, 2(or3) refers to the suckdown regions of Fig 2.13, and the multijet increment is given by

$$\frac{\Delta\Delta L_{s,2(\text{or}3)}}{T} = 0.001\left(\frac{h'-h}{D_e}\right)^j K_r \tag{2.68}$$

and the exponent j is given by

$$j = 2.2\left(\frac{e_{\text{ave}}}{D_e}\right)^{-0.5}\left(\frac{Y}{D_e}\right)^{0.22} E^{0.3}\text{NPR}^{-0.1} \tag{2.69}$$

and the value of Y used is the planform half width at the jet (or pair of jets).

Outboard of the jets and between the fountain and the lone jet in a three jet configuration, the suckdown is reduced to 80% of that for the two jet configuration:

$$\frac{\Delta\Delta L_{s,1(\text{or}4)}}{T} = 0.8\left(\frac{\Delta L_s}{T}\right)_{2-\text{jet}-\text{est}} \tag{2.70}$$

g. Effect of LIDs—two jets. Frequently strakes, or LIDs, are installed on the lower surface in an attempt to increase the fountain lift or minimize hot-gas ingestion. However the primary effect (Fig. 2.21) is not to increase the fountain positive pressures but 1) to reduce the suckdown pressures induced between the fountain and the jets and 2) to change pressures from negative to positive in some areas between the LIDs.

The lift increment caused by LIDs has been correlated with the fountain lift increment. For the two jet configurations the increment of lift caused by the LIDs is given by

$$\frac{\Delta L_{\text{LID}}}{T} = \frac{\Delta L_f}{T} K_{\text{LID}}\left(\frac{S_{\text{LID}}}{S_2 + S_f + S_3}\right)P_{\text{LID}} \tag{2.71}$$

where P_{LID} is the ratio of the perimeter closed by the LIDs to the total perimeter of the LIDs planform and K_{LID} is given by

$$K_{\text{LID}} = 12\left(\frac{h}{h_f}\right) - 10\left(\frac{h}{h_f}\right)^2 \tag{2.72}$$

Unfortunately the database for LID effects is even more limited than for fountain effects in general, and there are little data on such parameters as the effect of jet spacing; therefore, these expressions should be used with caution.

h. Effect of LIDs—three or four jets. If the fore or aft jets, or both, are replaced by a side-by-side pair, the lift increment caused by LIDs can be estimated by

$$\frac{\Delta L_{\text{LID}}}{T} = \frac{\Delta L_f}{T} K_{0,\text{LID}}\left(\frac{S_{\text{LID}}}{S_2 + S_f + S_3}\right)P_{\text{LID}}\left(\frac{S_{\text{LID}}}{S}\right)^{1.5} \tag{2.73}$$

where $K_{0,\mathrm{LID}}$ is given by

$$K_{0,\mathrm{LID}} = 2\left(\frac{h}{h'}E\right) - 1.7\left(\frac{h}{h'}E\right)^{2.5} \tag{2.74}$$

i. Pitching moment—two jet configurations. The net induced pitching moment is the sum of the moment increments caused by the lift or suckdown induced on each region:

$$\frac{\Delta M}{TD_e} = \frac{\Delta M_{s,1}}{TD_e} + \frac{\Delta M_{s,2}}{TD_e} + \frac{\Delta M_f}{TD_e} + \frac{\Delta M_{s,3}}{TD_e} + \frac{\Delta M_{s,4}}{TD_e} + \frac{\Delta M_{\mathrm{LID}}}{TD_e} \tag{2.75}$$

The moment contribution of each region is the lift increment multiplied by its effective arm. For two jet configurations the moment contribution of the fountain lift was found to be negligible, even for significant amounts of differential thrust.

The moment contribution of the regions between the fountain and the jets (regions $s,2$ and $s,3$) is the lift increment multiplied by the distance to the center of the suckdown area:

$$\frac{\Delta M_{s,2(\mathrm{or}3)}}{TD_e} = \frac{\Delta L_{s,2(\mathrm{or}3)}}{T}\frac{X_{\mathrm{e,inb}}}{D_e} \tag{2.76}$$

where, for configurations with little planform taper between the jets, the effective arm $X_{\mathrm{e,inb}}$ can be taken as

$$X_{\mathrm{e,inb}} = X_0 + \left(\frac{e - X_{\mathrm{o}}}{2}\right) \tag{2.77}$$

Outboard of the jets (regions $s,1$ and $s,4$) the effective arm is shorter than the distance to the center of the area and is given by

$$X_{\mathrm{e,out}} = e + 0.6\left(\frac{Y}{d}\right)^{0.25}\left(\frac{h}{D_e}\right)^{-0.15} X_{s,\mathrm{out}} \tag{2.78}$$

where Y is the half-width at the jet. The moment contribution of the outboard region is given by

$$\frac{\Delta M_{s,1(\mathrm{or}4)}}{TD_e} = \frac{\Delta L_{s,1(\mathrm{or}4)}}{T}\frac{X_{\mathrm{e,out}}}{D_e} \tag{2.79}$$

j. Pitching moment—three and four jets. With a side-by-side pair forward, or aft, positive pressures are generated at some distance from the moment reference point, and small fountain and LID moment increments are induced. The effective arm is given by

$$\frac{X_{f,m}}{D_e} = \frac{e_x}{D_e}\left(\frac{e_{y,f} - e_{y,r}}{e_{y,f} + e_{y,r}}\right)\left(\frac{h' - h}{h'}\right)^4 \tag{2.80}$$

The pitching-moment increment caused by the fountain is given by

$$\frac{\Delta M_f}{TD_e} = \frac{\Delta L_f}{T}\frac{X_{f,m}}{D_e} \tag{2.81}$$

and the pitching-moment increment caused by LIDs is given by

$$\frac{\Delta M_{\text{LID}}}{TD_e} = \frac{\Delta L_{\text{LID}}}{T}\frac{X_{f,m}}{D_e} \tag{2.82}$$

When the side-by-side pair is forward, the moment arm of the suckdown forward of the fountain is increased, and the moment contribution is given by

$$\frac{\Delta M_{s,2}}{TD_e} = \frac{\Delta L_{s,2}}{T}\frac{X_o + \frac{e_x - X_o}{2} + X_{f,m}/4}{D_e} \tag{2.83}$$

and the moment arm of the suckdown aft of the fountain is decreased and the moment contribution is given by

$$\frac{\Delta M_{s,3}}{TD_e} = \frac{\Delta L_{s,3}}{T}\frac{(X_{f,m}/4) - X_o - \left(\frac{e_x - X_o}{2}\right)}{D_e} \tag{2.84}$$

When the side-by-side pair is aft, the moment arm of the suckdown forward of the fountain is decreased, and the moment contribution is given by

$$\frac{\Delta M_{s,2}}{TD_e} = \frac{\Delta L_{s,2}}{T}\frac{X_o + \left(\frac{e_x - X_o}{2}\right) - X_{f,m}/4}{D_e} \tag{2.85}$$

and the moment arm of the suckdown aft of the fountain is increased and the moment contribution is given by

$$\frac{\Delta M_{s,1(\text{or}4)}}{TD_e} = \frac{\Delta L_{s,1(\text{or}4)}}{T}\left(\frac{e}{D_e} + 0.6\left(\frac{Y}{d}\right)^{0.25}\left(\frac{h}{D_e}\right)^{-0.15}\frac{X_{\text{s,out}}}{D_e}\right) \tag{2.86}$$

The moment contribution of the outboard suckdown regions is given by

$$\frac{\Delta M_{s,3}}{TD_e} = \frac{\Delta L_{s,3}}{T}\frac{-(X_{f,m}/4) - X_o - \left(\frac{e_x - X_o}{2}\right)}{D_e} \tag{2.87}$$

Nomenclature

A_j = total area of all jets, ft^2
b = exponent used in estimating width of fountain region [Eq. (2.40)]
C_p = pressure coefficient, based on the nominal jet dynamic pressure
$C_{\text{p,ave,f}}$ = average pressure in fountain region [Eqs. (2.42) and (2.46)]
$C_{\text{p,ave,s}}$ = average pressure in suckdown regions [Eqs. (2.48–2.51)]

D = diameter of plate, ft
D_e = effective diameter of total jet area, ft
D_p = effective diameter of total planform, ft
DP = decay parameter [Eq. (2.3)]
d = diameter of individual jet, ft
E = length/width ratio of multijet pattern [Eq. (2.62)]
e = half-distance between the jets, ft
e_{ave} = average jet spacing of multijet configurations [Eq. (2.61)], ft
e_y = half-width at front, or aft jet, whichever is greater, ft
exp = exponent used in single jet suckdown calculation [Eqs. (2.12) and (2.25)]
f = exponent used in estimating average fountain pressure [Eqs. (2.43) and (2.44)]
f_n = fineness ratio of the nozzle
f_p = fineness ratio of the planform
g = exponent used in estimating suckdown of two jet configurations [Eq. (2.59)]
H_s = height parameter used in estimating suckdown of two jet configurations [Eq. (2.52)]
h = height above the ground, ft
h_f = height of fountain for two jet configurations [Eq. (2.45)], ft
h_{tv} = height below which the trapped vortex develops [Eq. (2.8)], ft
h' = height of fountain for multijet configurations [Eq. (2.60)], (see Fig. 2.28), ft
h'' = intermediate height in estimating fountain lift for multijet configurations [Eq. (2.64)] (see Fig. 2.28), ft
i = exponent used in estimating suckdown of two jet configurations [Eq. (2.58)]
j = exponent used in estimating suckdown of multijet configurations [Eq. (2.69)]
K_{LID} = constant used in estimating effects of LIDs for two jet configurations [Eq. (2.72)]
$K_{n,f}$ = constant used in estimating effects of noncircular nozzles on fountain pressures [Eq. (2.33)]
$K_{n,s}$ = constant used in estimating effects of noncircular nozzles on suckdown [Eqs. (2.56) and (2.57)]
$K_{n,x}$ = constant used in estimating effects of noncircular nozzles on width of fountain region [Eq. (2.39)]
K_r = body shape factor [Eqs. (2.34) and (2.35)] (see Fig. 2.20)
K_s = constant used in estimating average suckdown pressure coefficient for multiple jet configurations [Eqs. (2.53–2.55)]
K_{sj} = constant used in estimating single jet suckdown [Eq. (2.11)]
$K_{0,\text{LID}}$ = constant used in estimating effects of LIDs for multijet configurations [Eq. (2.74)]
K_∞ = constant used in estimating out-of-ground-effect lift loss [see Eq. (2.5) and Fig. 2.1]
N = number of jets
per = perimeter of noncircular jet, ft

P_{LID} = ratio of perimeter enclosed by LIDs to the total planform perimeter
q_e = jet dynamic pressure at nozzle exit, lb/ft^2
q_z = jet dynamic pressure at distance z from nozzle exit, lb/ft^2
R = radius of plate, ft
r = radial station from jet center, or effective corner radius of body lower surface, ft
S = area of planform or region of planform under consideration, ft^2
S_{cs} = area of wing center section, ft^2
S_f = area affected by fountain pressure [Eq. (2.36)], ft^2
S_{LID} = area enclosed by LIDs, ft^2
S_w = wing area, ft^2
s = distance between jet centers, ft
$s_1\ s_2\ s_3\ s_4$ = suckdown regions (see Fig. 2.13)
T = total jet thrust, lb
T_l = trust of local jet, lb
T_0 = time zero (see Fig. 2.17), s
V_0 = freestream velocity, ft/s
w = half-width of body for jets outside the body (see Fig. 2.27) or half-wing chord for tandem jets in body when the wing chord at side of body does not extend from jet to jet, ft
x_c = distance from moment reference point to the geometric center of the total planform, ft
x_e = effective arm at which the lift increment acts to produce the pitching-moment increment [Eqs. (2.14) and (2.15)], ft (see Fig. 2.9)
$x_{f,m}$ = effective arm at which fountain lift acts for multiple jet configurations [Eq. (2.80)], ft
x_s = distance from the moment reference point to the geometric center of the area under consideration, ft
x_w = distance from moment reference point to the geometric center of the outer wing panels, ft
x_0 = half-width of the fountain region in the x direction [Eqs. (2.3), (2.38), and (2.41)] (see Fig. 2.27), ft
Y = half-width of the body at front or rear jet, ft
z = distance downstream from nozzle exit, ft
z_i = distance from nozzle at which maximum jet decay rate occurs (see Fig. 2.3b), ft
Δh = height of wing above nozzle exit, ft
ΔL = lift increment, lb
ΔM = pitching-moment increment, $\text{ft} \cdot \text{lb}$
Δp = pressure increment induced by jet, lb/ft^2
ΔS = area of suckdown region being considered, ft^2
ΔT = time increment, s
δ = jet deflection angle, measured from the horizontal, deg

Subscripts

aft = aft of jet
ave = average

b = body
cs = center section
dT = differential thrust
e = effective, or exit
f = fountain; or front jet
fwd = forward of jet
hw = high wing
inb = inboard
LID = lift improvement devices
out = outboard
p = planform
p,ave,f = average fountain pressure
r = rear jet; or body radius (see Fig. 2.20)
s,1 s,2 s,3 s,4 = suckdown regions (see Fig. 2.13)
w = wing
wb = wing body
δ = jet deflection angle
∞ = out-of-ground effect

References

[1]Cox, M., and Abbott, W. A., "Studies of the Flow Fields Created by Single Vertical Jets Directed Downwards Upon a Horizontal Surface," National Gas Turbine Establishment, Pyestock, Hants, England, NGTE Mem. No. M.390, Oct. 1964.

[2]Gentry, G. L., and Margason, R. J., "Jet-Induced Lift Losses on VTOL Configurations Hovering in and out of Ground Effect," NASA TND-3166, Feb. 1966.

[3]McLemore, H. C., "Jet-Induced Lift Loss of Jet VTOL Configurations in Hovering Condition," NASA TN D-3435, June 1966.

[4]Shumpert, P. K., and Tibbetts, J. G., "Model Tests of Jet-Induced Lift Effects on a VTOL Aircraft in Hover," NASA CR-1297, March 1969.

[5]Kuhlman, J. M., Ousterhout, D. S., and Warcup, R. W., "Experimental Investigation of Effects of Jet Decay Rate on Jet-Induced Pressures on a Flat Plate," NASA CR-158990, Nov. 1978.

[6]Kuhlman, J. M., and Warcup, R. W., "Experimental Investigation of Jet-Induced Loads on a Flat Plate in Hover out-of-Ground-Effect," NASA CR-159004, Feb. 1979.

[7]Margason, R. J., Arledge, T., Wardwell, D. A., Hange, C. E., and Naumowicz, T., "Jet Efflux Characteristics and Their Influence on STOVL Aircraft Propulsion-Induced Effects," Society of Automotive Engineers, Paper 962250, Nov. 1996.

[8]Margason, R. J., Arledge, T., Wardwell, D. A., Hange, C. E., and Naumowicz, T., "Influence of Jet Efflux Characteristics on STOVL Aircraft Propulsion-Induced and Ground Effects," *Royal Aeronautical Society Proceedings*, International Powered Lift Conference, London, Sept. 1998, pp. 25.1–25.13.

[9]Arledge, T. K., and Wardwell, D. A., "Swirl Effects on Jet Efflux Characteristics Relating to STOVL Aircraft Propulsion-Induced Effects," *Royal Aeronautical Society Proceedings*, International Powered Lift Conference, London, Sept. 1998, pp. 15.1–15.9.

[10]Kuhn, R. E., Bellavia, E. C., Corsiglia, V. R., and Wardwell, D. A., "On the Estimation of Jet-Induced Fountain Lift and Additional Suckdown in Hover for Two-Jet Configurations," NASA TM 102268, Aug. 1991.

[11]Bellavia, D. C., Wardwell, D. A., Corsiglia, V. R., and Kuhn, R. E., "Forces and Pressures Induced on Circular Plates by a Single Jet in Ground Effect," NASA TM 102816, March 1991.

[12]Kuhn, R. E., Bellavia, E. C., Wardwell, D. A., and Corsiglia, V. R., "On the Anomalies in Single-Jet Hover Suckdown Data," NASA TM 102261, Aug. 1991.

[13]Wyatt, L. A., "Static Test of Ground Effect on Planforms Fitted with a Centrally-Located Round Lifting Jet," U. K. Ministry of Aviation, CP 749, June 1962.

[14]Wardwell, D. A., Hange, C. E., Kuhn, R. E., and Stewart, V. R., "Jet Induced Ground Effects on a Parametric Flat Plate Model in Hover," NASA TM104001, Feb. 1993.

[15]Kuhn, R. E., Stewart, V. R., and Wardwell, D. A., "Estimation of Lift and Pitching Moment Induced on Jet STOVL Aircraft Hovering in Ground Effect," WL-TR-93-3046, Flight Dynamics Directorate, Wright Laboratory, Air Force Material Command, Wright Patterson Air Force Base, Ohio, Aug. 1993.

[16]Kuhn, R. E., Wardwell, D. A., and Blake, W. B., "A Method for Estimating the Lift and Pitching Moment on Jet STOVL Aircraft Hovering in Ground Effect," AIAA Paper 93-4816, International Powered Lift Conference, Santa Clara, CA, Dec. 1993, pp. 24–34.

[17]Hall, G. R., and Rogers, K. H., "Recirculation Effects Produced by a Pair of Heated Jets Impinging on a Ground Plane," NASA CR-1307, 1969.

[18]Kotansky, D. R., "Jet Flow Fields," AGARD, Rept. 710, '*Special Course on V/STOL Aerodynamics*,' April 1984, pp. 7-1–7-8.

[19]Saripalli, K. R., "Laser Doppler Velocimeter Measurements in a 3-D Impinging Twin-Jet Fountain Flow," NASA CP 2462 *Proceedings of the 1985 NASA Ames Research Center's Ground Effects Workshop*, Aug. 1985, pp. 147–160.

[20]Wardwell, D. A., Bellavia, D. C., Corsiglia, V. R., and Kuhn, R. E., "Dynamic Response of Induced Pressures, Suckdown and Temperatures for Two Tandem Jet STOVL Configurations," NASA TM 103934, July 1992.

[21]Cabrita, P. M., Saddington, A. J., and Knowles, K., "Unsteady Features of Twin-Jet STOVL Ground Effects," AIAA 2002-6014, Nov. 2002.

[22]Kuhn, R. E., "An Engineering Method for Estimating the Induced Lift on V/STOL Aircraft Hovering in and out of Ground Effect," NADC Rept. NADC-80246-60, Naval Air Development Center, Warminster, PA, Jan. 1989.

[23]Spong, E. D., Kamman, J. H., and Flood, J. D., "V/STOL Jet-Induced Interactions," McDonnell Aircraft Co., *Proceedings of Workshop on V/STOL Aerodynamics*, Naval Postgraduate School, Monterey, CA, 1979, pp. 490–508.

[24]Wohllebe, F. B., and Migdal, D., "Some Basic Test Results of V/STOL Jet-Induced Lift Effects in Hover," AIAA Paper 70-0339, Jan. 1979.

[25]Foley, W. H., and Sansome, J. A., "V/STOL Propulsion-Induced Aerodynamics Hover Calculation Method," General Dynamics, NADC-78242-60, Naval Air Development Center, Warminster, PA, Feb. 1966.

[26]Bellavia, D. C., Wardwell, D. A., Corsiglia, V. R., and Kuhn, R. E., "Suckdown, Fountain Lift and Pressures Induced on Several Tandem Jet V/STOL Configurations," NASA TM 102817, March 1991.

[27]Hange, C. E., and Wardwell, D. A., "Small Scale Ground Effects and Hot Gas Ingestion Research at NASA Ames," AIAA Paper 92-4252, Aug. 1992.

[28]Vogler, R. D., "Interference Effects of Single and Multiple Round and Slotted Jets on a VTOL Model in Transition," NASA TN D-2380, Aug. 1964

[29]Levin, D. B., and Wardwell, D. A., "Single Jet-Induced Effects on Small-Scale Hover Data in Ground Effect," AIAA *Journal of Aircraft*, Vol. 34, No. 3, 1997, pp. 400–407.

[30]Ing, D. N., and Zhang, X., "An Experimental and Numerical Study of Single Jet Impingement Ground-Effect Lift Loss," AIAA Paper 93-4887, Dec. 1993; also *1993 AIAA International Powered Lift Conference*, Santa Clara, CA, Dec. 1993, pp. 293–303.

[31]Shih, C., Lou, H., and Alvi, F., "Microjet Control of Supersonic Impinging Jets—Control Strategy and Physical Mechanisms," AIAA Paper 2002-6009, Nov. 2002.

Chapter 3

Transition Out-of-Ground Effect

I. Introduction

IN THE transition between hover and conventional flight, the streams from the lifting jets, or fans, are swept back by the freestream and rolled up into vortex pairs (Fig. 3.1).

These vortices, along with the entrainment action and blockage effect of the jets, induce suction pressures on the bottom of the configuration beside and behind the jets and a smaller region of positive pressures ahead of the jet. These induced pressures generally produce a lift loss and a nose-up moment. The severity of these effects depends on the jet arrangement, aircraft configuration, and on the aircraft flight speed.

II. Jet/Freestream Interaction

Investigations of the jet/freestream interaction have been the subject of many investigations (for example, see Refs. 1–19). As pointed out in Ref. 1, work on understanding a jet in a crossflow preceded interest in V/STOL aircraft and ranged from smokestack plumes to fuel injection and air jets used to improve furnace combustion.

Emphasis relative to V/STOL aircraft has included experimental studies, flow visualizations, and the development of methods for predicting the jet path and the pressures induced on the lower surface of the aircraft. This work has shown that in addition to the roll up of the jet into a vortex pair there is a turbulent region in the wake of the jet and a horseshoe vortex looped around the jet in the exit plane (Fig. 3.2).

The freestream is entrained into the jet and into the turbulent wake behind it. As depicted in Fig. 3.3 (Ref. 13), particles near the surface are entrained into the jet and rolled up into the jet vortices or are sucked into the wake and from there back toward the jet.

Figure 3.4 shows the particle traces near the surface as well as the flow in the plane of symmetry and more clearly shows the flow being drawn into the jet from the rear and the apparent stagnation point on the surface a few diameters behind the jet.

The appropriate parameter for correlating the effects of freestream velocity with jets of various pressure ratios and temperatures was shown, in Ref. 2, to be the effective velocity ratio V_e (Fig. 3.5).

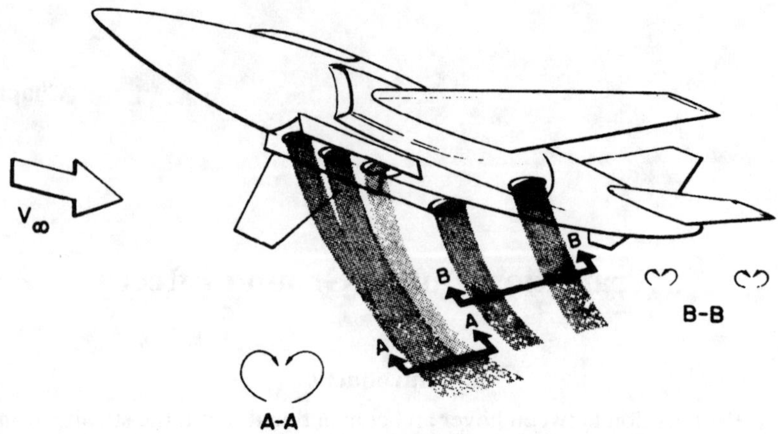

Fig. 3.1 Jet/freestream interaction in transition.

The effective velocity ratio V_e is defined as the square root of the ratio of free-stream dynamic pressure q_o to jet dynamic pressure q_j:

$$V_e \equiv \sqrt{\frac{q_o}{q_j}} \tag{3.1}$$

Other results of the studies of jet path and the pressures induced included the finding that the induced suckdown pressures, and by inference the lift loss,

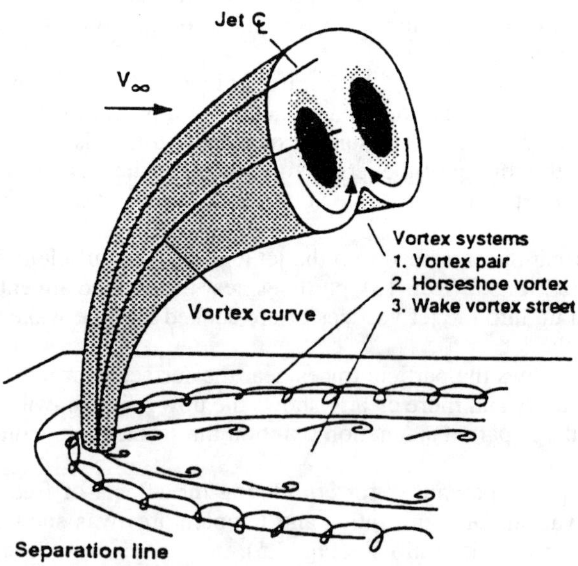

Fig. 3.2 Schematic of a jet in a crossflow.

Fig. 3.3 Particle traces in the jet and near the surface for $V_e = 0.17$.

can be reduced by switching from round to elongated slot nozzles (Fig. 3.6) and that with two jets in tandem (Fig. 3.7) the aft jet is not deflected aft as much as the forward jet because it is operating in the reduced velocity wake of the front jet.

In most cases the jet wake-induced pressures result in a nose-up moment and a net loss in lift on the configuration (Fig. 3.8). The induced effects increase rapidly with aircraft velocity and are a function of the ratio of planform to jet area ratio and the location of the jet(s).

Although for most configurations the jet-induced flows cause a lift loss, it has been found[3] that if the jets are moved aft to a position near or aft of the wing trailing edge a favorable lift increment is induced (Fig. 3.9). Apparently, the jet at or near the trailing edge of the wing is acting like a small-span jet flap.

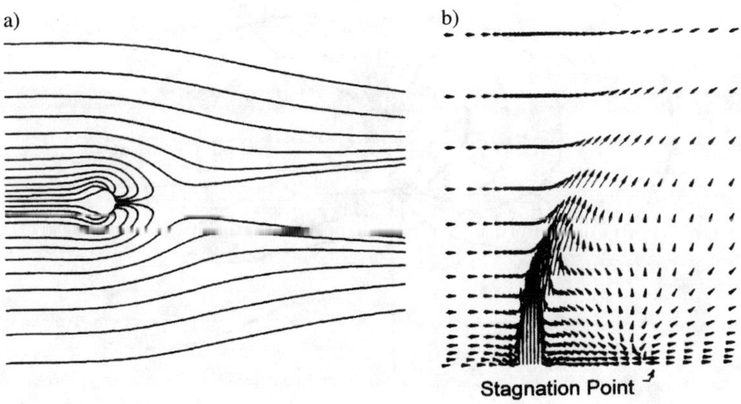

Fig. 3.4 a) Typical particle traces on the surface. b) Velocity vector field in the symmetry plane (right) (from Ref. 13).

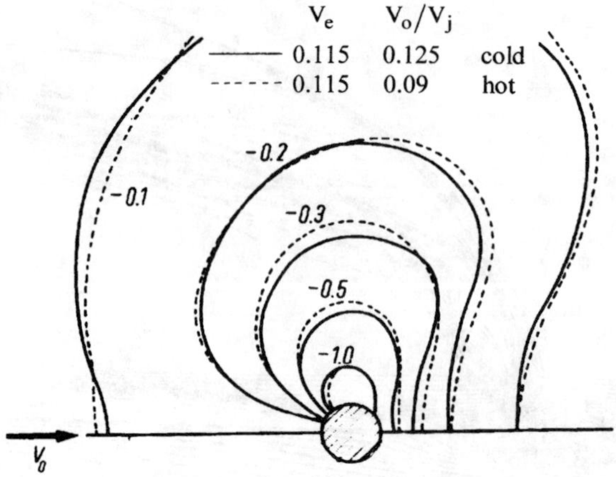

Fig. 3.5 Correlation of surface pressures induced by cold and hot jets.[2]

When McDonnell Douglas started the development of the AV-8B (Fig. 1.1), they were aware of effects presented in Ref. 3 and the advantages of having at least part of the jet thrust near the wing trailing edge. The design of the engine and balance considerations precluded moving the jets aft, but because the program was to include a new wing they designed a new large chord flap that, in effect, moved the wing trailing edge (when the flap was deflected) closer to the rear nozzle. The lift loss induced by the front nozzles was reduced by extending the wing leading edge, but, as shown in Fig. 3.10, the favorable lift induced by

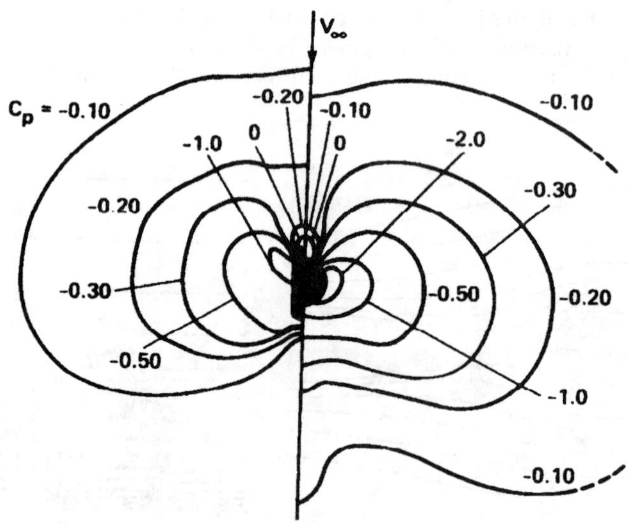

Fig. 3.6 Comparison of pressures induced by a longitudinal slot jet (left) to those induced by a circular nozzle (right).[14]

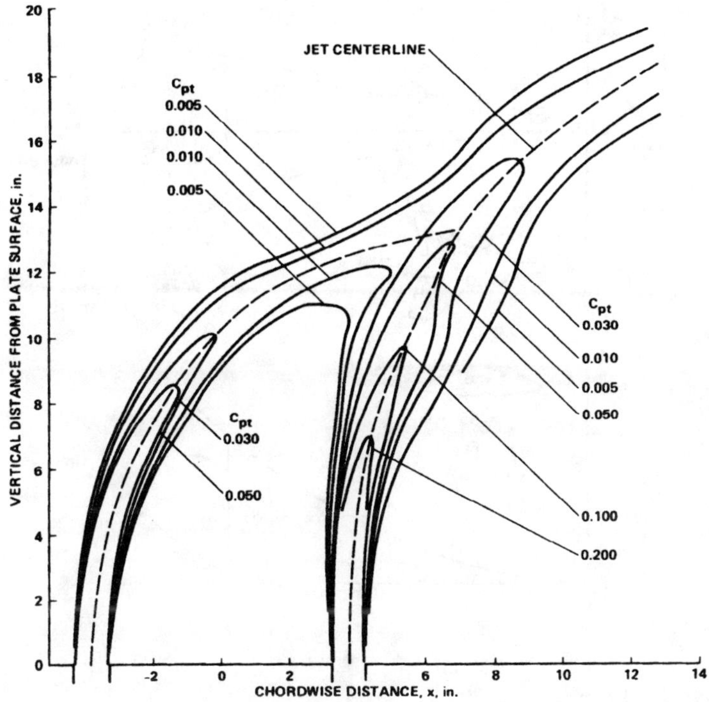

Fig. 3.7 Paths taken by two jets in tandem.[15]

the rear jets near the flap balances out the lift loss of the front jets and gives the configuration a net-induced lift increment near zero, greatly increasing the overload STOL capability of the AV-8B over that of the AV-8A.

During the Joint Strike Fighter (JSF) competitive phase in the 1990s, the McDonnell Douglas Company developed a fighter configuration that used deflected convergent jets with a supercritical pressure ratio. The experimental jet-induced effects data did not demonstrate correlation using the effective velocity ratio V_e. These jets differ from subcritical and critical pressure ratio jets. The convergent, supercritical jet has an underexpanded plume at the jet exit that effectively increases the jet exit area and enlarges the plume geometry. Reference 20 evaluates supercritical jet-induced effects using data from this aircraft configuration to identify a more appropriate correlation parameter. New definitions of a supercritical effective velocity ratio V_e^* and of a supercritical effective diameter D_e^* that are based on the geometry of an underexpanded supercritical pressure ratio convergent jet plume were developed. Several encouraging examples of its application to the 12% small-scale powered model (SSPM) unpublished data obtained in the Boeing Mini-Speed Wind Tunnel were presented. This approach was successfully used to correlate surface-pressure data located downstream of the right rear jet and for the jet-induced lift and pitching moment increments caused by the rear jets alone. Additional complete

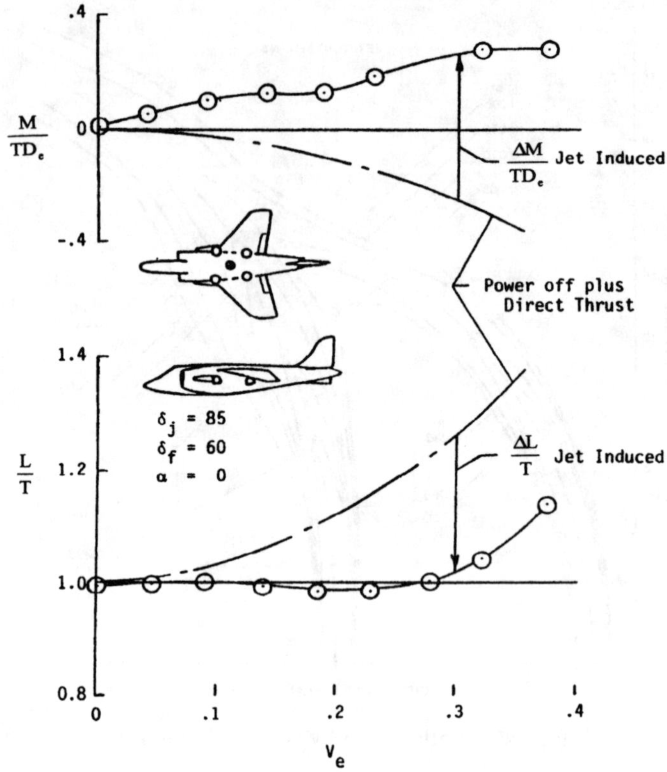

Fig. 3.8 Typical jet-induced effect on lift and pitching moments.

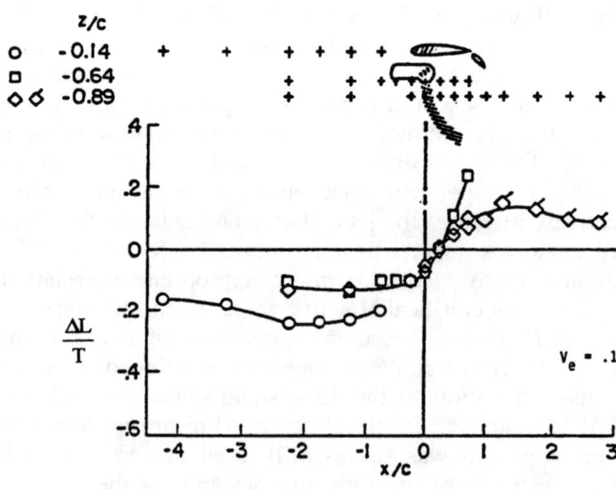

Fig. 3.9 Effect of longitudinal position of jet on induced lift.[3]

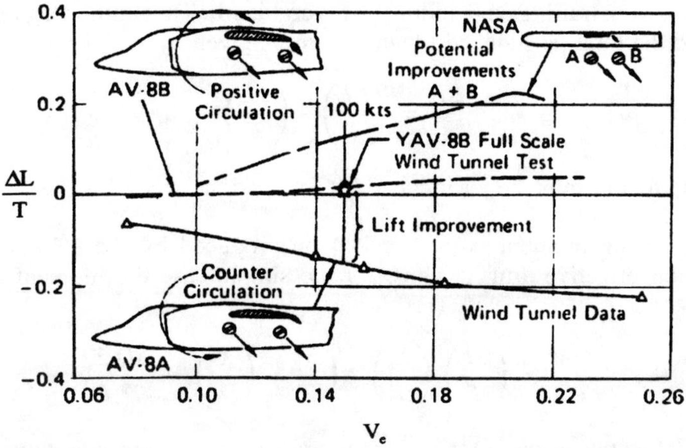

Fig. 3.10 Effect of wing leading edge and flap relocation in reducing the lift loss for the AV-8B (from Ref. 17).

configuration data correlations for induced-lift increment were also encouraging. Further evaluation is needed to fully validate this approach.

III. Induced Lift, Drag, and Moment

The total forces and moments on a V/STOL configuration in transition can be considered to be made up of the sum of the aerodynamic forces (based on power off data), the components of the direct thrust, and the induced increments:

$$\text{Lift} = C_L q_o S + T \sin(\alpha + \delta_j) + T\left(\frac{\Delta L_\infty + \Delta L_t}{T}\right) \tag{3.2}$$

$$\text{Drag} = C_D q_o S + T \cos(\alpha + \delta_j) + T\left(\frac{\Delta D}{T}\right) \tag{3.3}$$

$$\text{Moment} = C_m q_o S\bar{c} + T \sin(\alpha + \delta_j)X + TD_e\left(\frac{\Delta M}{TD_e}\right) \tag{3.4}$$

where T is the total thrust at the deflection angle δ_j.

The lift increment ΔL_∞ is the small lift loss induced out-of-ground effect, from Sec. I in Chapter 2, and ΔL_t is the lift loss induced in transition estimated by the method presented next along with the methods for estimating the drag ΔD and moment ΔM increments.

The method used here for estimating the jet-induced effects in transition is based on the method presented in Refs. 21 and 22. The method was initially presented in Sec. 2.2.2 of Ref. 21 in terms of lift/thrust ratios in graphical form and was converted to equations in coefficient form in Sec. 4.1.3.3 of Ref. 22.

As already indicated, a single jet located near the centroid of the planform induces a lift loss along with a nose-up moment, whereas a jet located near or behind the wing trailing edge induces a favorable lift. The method accommodates both, and the total jet-induced lift increment is given by

$$\frac{\Delta L_t}{T} = \left(\frac{\Delta L}{T}\right)_j + \left(\frac{\Delta L}{T}\right)_\Gamma \tag{3.5}$$

where the first term is the jet-induced lift loss and the second is the "jet flap" lift gain term.

The pitching moment is assumed to be produced by these lift increments acting their effective arm X. Thus, for the simple case the moment increment is given by

$$\frac{\Delta M}{TD_e} = \left(\frac{\Delta L}{T}\right)_j \left(\frac{X}{D_e}\right)_j + \left(\frac{\Delta L}{T}\right)_\Gamma \left(\frac{X}{D_e}\right)_\Gamma + \frac{\Delta D}{T}\frac{Z_i}{D_e} \tag{3.6}$$

where the first two terms are the moment produced by the preceding two lift terms and the third term is the moment produced by the inlet momentum drag.

In the method presented here the induced lift is estimated for the individual jet (or side-by-side pair), and the increments for multiple jets are summed.

A. Lift Loss

The primary factors that determine the lift loss induced by the jet/freestream interaction are the ratio of the area surrounding the jet to the jet area and the crossflow velocity as expressed by the effective velocity ratio V_e. The basic lift loss (Fig. 3.11) for a vertical jet at the center of a square plate is calculated by

$$\left(\frac{\Delta L}{T}\right)_{\text{j,basic}} = 35V_e^{5.5}\left(\sqrt{\frac{S}{A_j}} - 1\right) - 3V_e^2\left(\sqrt{\frac{S}{A_j}} - 1\right)^{\frac{2}{3}} \tag{3.7}$$

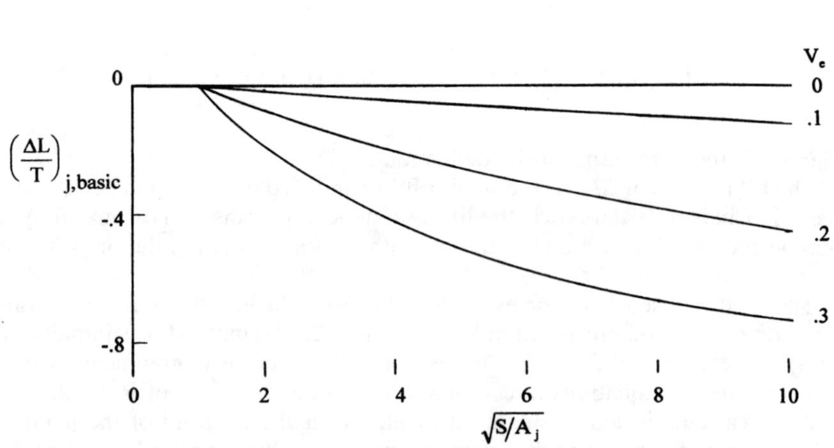

Fig. 3.11 Basic lift loss induced by a vertical jet at the center of a square plate.

The effects of such factors as planform aspect ratio, jet position, nozzle shape, and deflection are accounted for by multiplying the basic lift loss by adjustment factors:

$$\left(\frac{\Delta L}{T}\right)_j = \left(\frac{\Delta L}{T}\right)_{\text{j,basic}} (K_{L,A})(K_{L,x})(K_{L,z})(K_{L,y})(K_{L,y'})(K_{L,n})\frac{\delta_j}{90} \tag{3.8}$$

The constant $K_{L,A}$ corrects for the aspect ratio of the planform:
For bodies (i.e., $A \leq 1.0$):

$$K_{L,A} = A^{0.05} - 0.5(1 - A)^{3.6} \tag{3.9}$$

for wings (i.e., $A \geq 1.0$):

$$K_{L,A} = 1 - 0.4[1 - (1/A)]^4 \tag{3.10}$$

The constant $K_{L,x}$ corrects for the longitudinal position of the jet:
For bodies:

$$K_{L,x} = 1.35 - 0.7(S_f/S) - 2.5[(S_f/S) - 0.5]^{2.5} \tag{3.11}$$

where S_f is the body area ahead of the jet:
For wings (Fig. 3.12), with the jet ahead of $\bar{c}/2$:

$$K_{L,x} = 1.0 \tag{3.12}$$

For wings, with the jet between $\bar{c}/2$ and the trailing edge:

$$K_{L,x} = 1 - \{2[(x_j/c_j) - 0.5]\}^2 \tag{3.13}$$

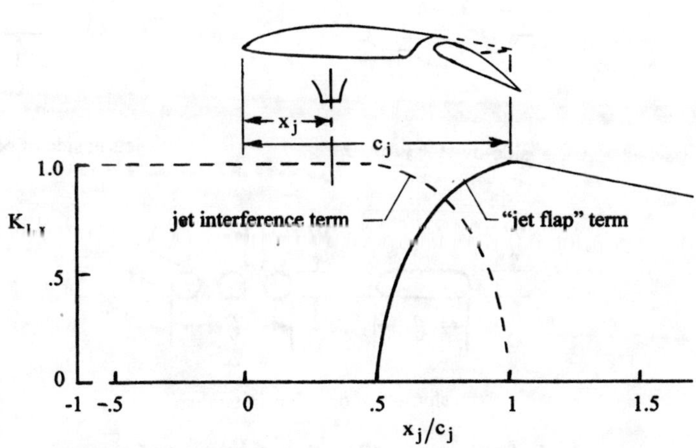

Fig. 3.12 Factor for adjusting for longitudinal position of the jet.

When the jet is aft of the midchord point $c_j/2$, a lift gain is experienced, which is estimated by the modified jet flap approach discussed in the lift gain section [Eqs. 3.25–3.32].

The constant $K_{L,z}$ corrects for the vertical distance that the jet exit extends below the lower surface:

$$K_{L,z} = 1 - 0.15\left(\frac{z}{d}\right) \tag{3.14}$$

If the single jet is replaced by a side-by-side pair, the lift loss is increased if the distance between them is less than three diameters. The constant $K_{L,y}$ corrects for the lateral spacing of the jets (see Fig. 3.13).

For values of y/d between 1 and 3,

$$K_{L,y} = 1.2 - 0.1\left(\frac{y}{d} - 1\right) \tag{3.15}$$

For values of y/d greater than three and for single jet configurations,

$$K_{L,y} = 1.0 \tag{3.16}$$

(Note that values of y/d less than 1.0 cannot exist; the jets would be overlapped.) If the jets are near the side of the body, or outside the body (see Fig. 3.13), the lift loss is reduced by the factor $K_{L,y'}$.

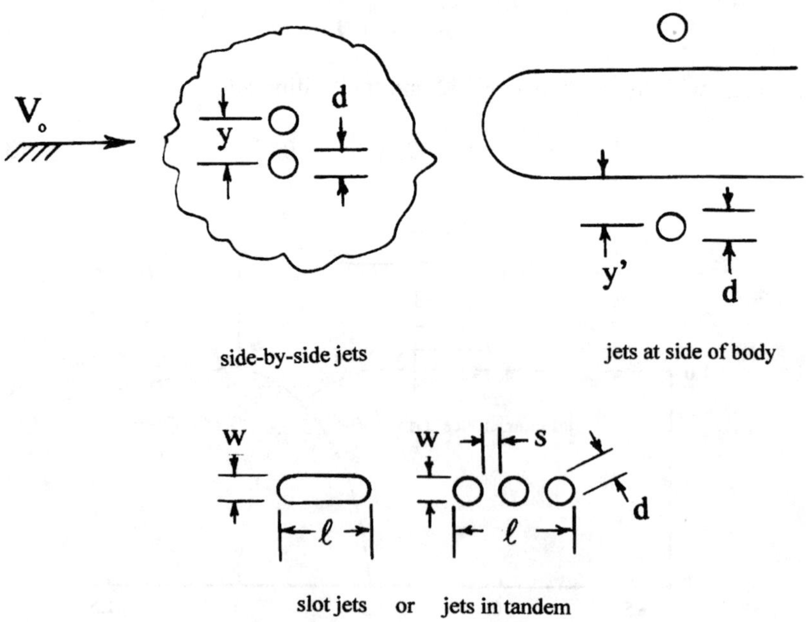

Fig. 3.13 Definitions of jet location, spacing, and geometry.

For values of y'/d less than 0.353,

$$K_{L,y'} = 1.0 \tag{3.17}$$

For values of y'/d between 0.353 and 2.0,

$$K_{L,y'} = 0.85 - 0.425(y'/d) \tag{3.18}$$

For values of y'/d greater than 2.0,

$$K_{L,y'} = 0 \tag{3.19}$$

If jets near either side of the body are splayed outward away from each other, their induced effect will be reduced, similar to increasing their spacing. The effects of splay can be estimated by using the lateral distance from the side of the body at the point on the projected centerline of the jet three diameters downstream of the nozzle, that is, for jets splayed outward,

$$y'_{\phi} = y'_{\phi=0} + 3d(\sin\,\phi) \tag{3.20}$$

If the jet is noncircular or several jets are closely spaced in tandem [that is, if s/d is less than 1.0 (see Fig. 3.13)], the reduction in lift loss can be estimated by the factor $K_{L,n}$, which is given by

$$K_{L,n} = 1 - (14V_e^{1.5} - 22V_e^{2.5})\left[1 - \left(\frac{\mathrm{W}}{\ell}\right)^{0.3}\right] \tag{3.21}$$

For more widely spaced tandem jets, the downstream jet is operating in the reduced velocity wake of the upstream jet, and their induced effects are reduced. The lift-loss increments are estimated for the front and rear jets separately and summed. The effective velocity ratio at the rear jet is calculated as suggested by Ref. 5 by

$$V_{e,\text{rear}} = V_{e,\text{front}}\frac{(s/d_f) - 1}{(s/d_f) + 0.75} \tag{3.22}$$

The induced effects caused by each jet are calculated as already shown for single jets using the thrust, velocity ratio, and ratio of planform to individual jet area appropriate to each:

$$\sqrt{\frac{S}{A_{j,\text{front}}}} \quad \text{and} \quad \sqrt{\frac{S}{A_{j,\text{rear}}}}$$

and the lift losses are summed using thrust weighting:

$$\frac{\Delta L}{T} = \left(\frac{\Delta L}{T}\right)_{\text{front}} \left(\frac{T_{\text{front}}}{T}\right) + \left(\frac{\Delta L}{T}\right)_{\text{rear}} \left(\frac{T_{\text{rear}}}{T}\right) \tag{3.23}$$

For four jet configurations with side-by-side pairs fore and aft, the induced effects of each pair are estimated separately and summed. The induced effects of the rear pair are estimated using the reduced velocity ratio at the rear jets based on the longitudinal spacing and individual diameters of the front jets.

For wing body configurations the body and wing increments are calculated separately and summed:

$$\left(\frac{\Delta L}{T}\right)_j = \left(\frac{\Delta L}{T}\right)_{j,\text{body}} + \left[\left(\frac{\Delta L}{T}\right)_{j,\text{wing}} - \left(\frac{\Delta L}{T}\right)_{j,\text{cs}}\right]_{\text{exposed-wing}} \tag{3.24}$$

where the wing center section cs is calculated as a low-aspect-ratio surface and subtracted from the calculation for the complete wing to adjust for body overlap. For high wing configurations the calculation for both the wing and center section are adjusted for wing height by the factor $K_{L,z}$.

B. Lift Gain

A lift gain is experienced when the jets are located at, near, or aft of the wing trailing edge (Fig. 3.9). This lift gain is caused by a jet flap action of the jets in augmenting the lifting circulation of wing. The lift increments induced are a function of the location of the jet (Fig. 3.9), reaching a peak when the jet is located at the wing, or flap, trailing edge. They are also much smaller than those normally associated with jet flaps because the spanwise extent of the wing affected is much smaller.

Because of the low effective aspect ratio of the body, the lift induced on the body is very small and is considered negligible in the method presented here. A lift gain is computed on the wing even for configurations with the jet(s) located in the body.

The basic lift gain term $(\Delta L/T)_{\Gamma,\text{basic}}$ (Fig. 3.14) is based on the jet flap theory of Ref. 23 and corresponds to a full-span jet directed vertically downward at the wing/flap trailing edge. This basic lift term was obtained by a conversion and replotting of the jet flap theory in terms of the lift-thrust ratio ($\Delta L_\Gamma/T = \Delta C_{L,\Gamma}/C_\mu$) and a parameter combining the effective velocity ratio and the area ratio ($V_e\sqrt{S/2A_j} = 1/\sqrt{C_\mu}$). The later term is derived for, and applies strictly only to, subsonic jets. The extent to which the methods presented here apply to higher nozzle-pressure-ratio supercritical jets are unknown.

The conversion of the jet flap theory[23] in terms of lift/thrust ratio produced an almost straight-line variation of $\Delta L/T$ vs $V_e\sqrt{S/2A_j}$. Therefore for the purposes of the present method, it is simplified to

$$\left(\frac{\Delta L}{T}\right)_{\Gamma,\text{basic}} = 2A_w^{0.36} V_e \sqrt{\frac{S}{2A_j}} \tag{3.25}$$

where A_w is the wing aspect ratio.

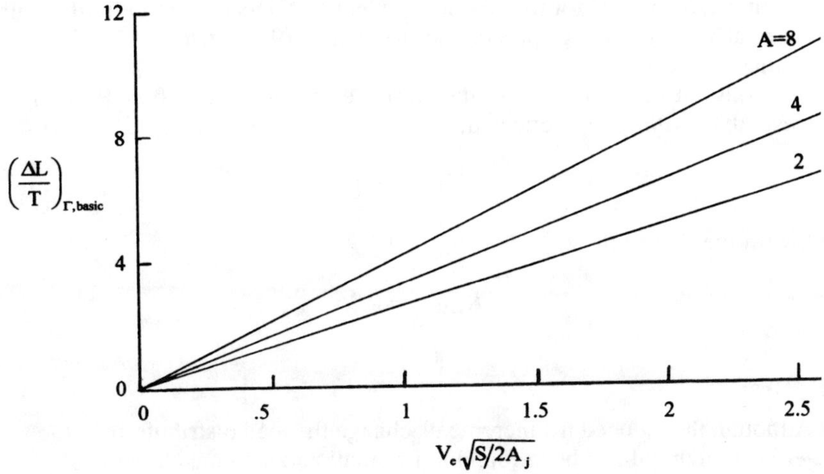

Fig. 3.14 Basic jet-flap-induced lift term.

The effects of such factors as jet position, jet deflection, the percentage of wing span affected, and the presence or absence of a flap are accounted for by multiplying the basic lift gain by adjustment factors:

$$\left(\frac{\Delta L}{T}\right)_{\Gamma} = \left(\frac{\Delta L}{T}\right)_{\Gamma,\text{basic}} K_{L,x} K_{L,z} \frac{Nd}{b} \frac{\delta_j}{90} K_{L,\text{flap}} \tag{3.26}$$

The constant $K_{L,x}$ corrects for the effect of the longitudinal location of the jet (Fig. 3.12). Ahead of the wing midchord point, the jet-flap-induced lift is negligible, and the correction factor is given by

$$K_{L,x} = 0 \tag{3.27}$$

The maximum jet-flap-induced lift increment occurs when the jet is at the wing trailing edge. Between the midchord point and the trailing edge:

$$K_{L,x} = 1 - \left[2\left(1 - \frac{x_j}{c_j}\right)\right]^2 \tag{3.28}$$

and aft of the wing trailing edge:

$$K_{L,x} = 1 - 0.25\left(\frac{x_c}{\bar{c}} - 1\right) \tag{3.29}$$

The constant $K_{L,z}$ corrects for the vertical distance that the jet exit extends below the lower surface:

$$K_{L,z} = 1 - 0.15\left(\frac{z}{d}\right) \tag{3.30}$$

The term Nd/b corrects for the spanwise extent of N jets having individual diameter d relative to the wing span b, and the term $\delta_j/90$ adjusts for jet deflections other than vertical.

The constant $K_{\mathrm{L,flap}}$ adjusts for the presence or absence of a flap. Reference 21 suggests that, with a flap deflected,

$$K_{L,\mathrm{flap}} = 1 \tag{3.31}$$

and, with the flap retracted,

$$K_{L,\mathrm{flap}} = 0.25 \tag{3.32}$$

C. Drag

Although the induced lift increments change the load distribution on the wing, the effects on drag have been found to be small and are considered negligible in the present method.

However, for aircraft, or models, with operating inlets a significant drag is created by the inlet momentum drag. This drag is the inlet mass flow multiplied by the transitional velocity. In some cases, particularly model tests with operational inlets in which the inlets and exits are powered separately, the inlet and exit weight flows differ. In these cases the drag is given by

$$\frac{\Delta D_i}{T} = \frac{w_i}{w_j} V_e \tag{3.33}$$

D. Pitching Moment (Bodies)

The jet-induced pitching moment is expressed as the sum of the jet-induced lift increments multiplied by their effective moment arms. For a body with a single jet located at the moment reference point, the pitching moment caused by the jet-induced lift loss is given by

$$\frac{\Delta M}{TD_e} = \left(\frac{\Delta L}{T}\right)_j \left(\frac{X}{D_e}\right)_j \tag{3.34}$$

and the moment arm is given by

$$\left(\frac{X}{D_e}\right)_j = \left(\frac{X}{D_e}\right)_{j,\mathrm{basic}} (K_{m,A})(K_{m,x})(K_{m,n})(K_{m,y}) \tag{3.35}$$

where the basic arm for a single circular jet located at the center of a square plate (which produces an induced pressure distribution similar to that shown in Fig. 3.5), and the effective moment arm is given by

$$\left(\frac{X}{D_\mathrm{e}}\right)_{j,\mathrm{basic}} = -0.2\sqrt{\frac{S}{A_j}} \tag{3.36}$$

The constant $K_{m,A}$ corrects for the aspect ratio of the planform:

$$K_{m,A} = K_1 + (K_1 - 1)\frac{1 - \sqrt{S_b/A_j}}{A\sqrt{S_b/A_j}} \tag{3.37}$$

where

$$K_1 = 1.307 - 0.37A + 0.063A^{1.67} \tag{3.38}$$

The constant $K_{m,x}$ corrects for the longitudinal position of the jet as follows:
For jets forward of the body midpoint $[(S_f/S) \leq 0.5]$:

$$K_{m,x} = 1 + \left[0.22 - 0.44\left(\frac{S_f}{S}\right) - \left|\frac{S_f}{S} - 0.5\right|^3\right]\frac{10}{\sqrt{S/A_j}} \tag{3.39}$$

And for jets aft of the body midpoint:

$$K_{m,x} = 1 + \left[0.22 - 0.44\frac{S_f}{S} - 0.7\left(\frac{S_f}{S} - 0.5\right)^2\right]\frac{10}{\sqrt{S/A_j}} \tag{3.40}$$

The constant $K_{m,n}$ corrects for noncircular nozzle:

$$K_{m,n} = \sqrt{\frac{w}{l}} \tag{3.41}$$

And the constant $K_{m,y}$ corrects for lateral spacing of a side-by-side pair:
For values of y/d between 1 and 2:

$$K_{m,y} = 1.8 - 0.8\left(\frac{y}{d} - 1\right) \tag{3.42}$$

And for values of y/d greater than 2 and for single jet configuration:

$$K_{m,y} = 1.0 \tag{3.43}$$

For bodies with multiple jets, or other cases where the jet is not located at the moment reference point, the distance from the moment reference point to the center of the jet must, of course, be added to the moment arm already calculated:

$$\left(\frac{\Delta M}{TD_e}\right)_{\text{body}} = \left(\frac{\Delta L_1}{T_1}\right)\left(\frac{T_1}{T}\right)\left(\frac{X_1 + X_1'}{D_e}\right) + \left(\frac{\Delta L_2}{T_2}\right)\left(\frac{T_2}{T}\right)\left(\frac{X_2 + X_2'}{D_e}\right) + \cdots \tag{3.44}$$

where X' is positive when the jet is ahead of the moment reference point.

E. Pitching Moment (Wings)

The lift loss induced on the wing produces an angle of attach type load distribution[3] with the center of aerodynamic load at the quarter-chord point. The moment arm of the jet-induced lift loss on the wing is therefore taken as the distance from the moment reference point to the quarter-chord of the mean geometric chord of the exposed wing:

$$\left(\frac{\Delta M}{TD_e}\right)_{j,\text{wing}} = \left(\frac{\Delta L}{T}\right)_{j,\text{wing}} \left(\frac{X_{\text{j}}}{D_{\text{e}}}\right) \tag{3.45}$$

where the distance from the moment reference point to the quarter-chord point X_j is positive if the quarter-chord point is ahead of the moment reference point.

The jet flap effect, on the other hand, produces a camber-type load distribution with the center of lift at the midchord point. The moment arm of the jet flap lift gain on the wing is therefore taken as the distance from the moment reference point to the midpoint of the mean geometric chord of the exposed wing:

$$\left(\frac{\Delta M}{TD_e}\right)_{\Gamma,\text{wing}} = \left(\frac{\Delta L}{T}\right)_{\Gamma,\text{wing}} \left(\frac{X_\Gamma}{D_e}\right) \tag{3.46}$$

where the distance from the moment reference point to the midchord point X_Γ is negative if the midchord point is aft of the moment reference point.

F. Pitching Moment (Inlet Drag)

An additional pitching-moment increment is generated by the effect of turning the inlet flow into the jet engine or fan inlets (Fig. 3.15). This moment increment is small relative to the increment induced by the exit flow, but it can be estimated by multiplying the inlet momentum drag [Eq. (3.32)] by the effective arm. For forward-facing inlets this arm is the vertical distance from the moment reference point to the center of the inlet.

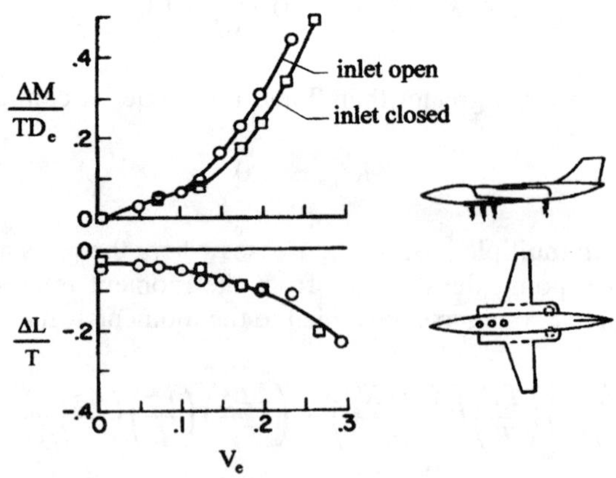

Fig. 3.15 Effect of inlet flow on pitching moments.[24]

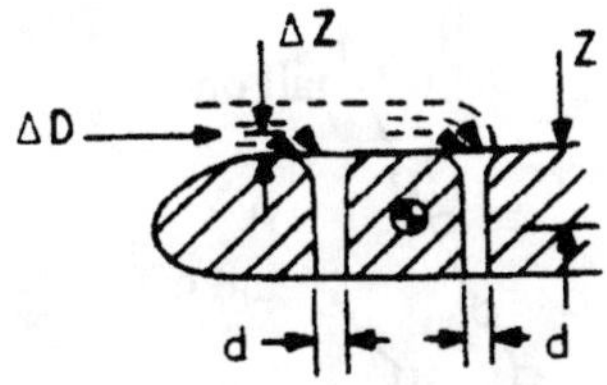

Fig. 3.16 Effective moment arm of upper-surface inlets.

For upper-surface inlets the moment arises from the high suction pressures generated on the upstream side of the inlet, and correspondingly lower pressures on the downstream side, as a result of turning the inlet flow into the inlet. The moment can be expressed, however, as the inlet drag multiplied by an effective arm, which includes an incremental distance ΔZ above the upper surface as depicted in Fig. 3.16.
And the moment caused by inlet momentum drag ΔD_i is given by

$$\left(\frac{\Delta M}{D_e}\right)_i = \frac{\Delta D_i}{T}\frac{Z_i + \Delta Z}{D_e} \tag{3.47}$$

There have been few investigations of the factors that determine this increment of effective arm, but the little data that exist suggest that it is approximated by

$$\Delta Z = 0.75d \tag{3.48}$$

IV. Downwash at the Tail

In addition to inducing suction pressures on the lower surface of the aircraft, the vortices generated in the jet wake can induce additional downwash at the tail. The pitching-moment increment caused by this downwash can be as large as that caused by the moment increment induced on the wing body with the tail off (Fig. 3.17). In this case the additional increment of pitching moment corresponds to a downwash angle of about 5 deg.

There have been few investigations of the jet-induced downwash. The effect of tail height, from one study,[7,8] is presented in Fig. 3.18. As would be expected the downwash angle increases as hovering is approached (as the effective velocity ratio V_e is reduced) because of the increasing jet strength relative to the freestream. The downwash angle also increases as the tail is lowered because the tail is getting closer to the vortices in the jet wake.

The effect of jet deflection on the jet-induced downwash, on a Harrier-type configuration, is shown in Fig. 3.19. The higher downwash for the lower jet deflection is probably caused by the vortices in the jet wake being moved closer to the tail by the jet being deflected aft. This configuration also shows the surprising result of the downwash decreasing as the effective velocity ratio is reduced. In this case the wing was carrying lift in the power-off case, as indicated by the level of power-off downwash shown. The jet-induced lift loss reduces the net lift, and, because this induced left loss is concentrated inboard,

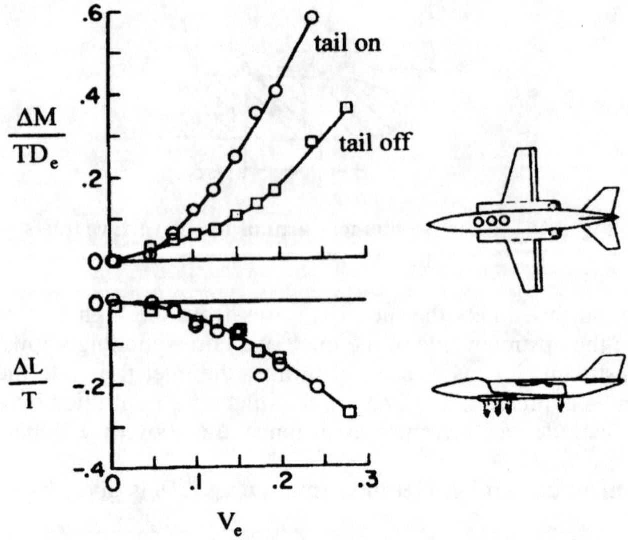

Fig. 3.17 Pitching moment caused by jet-induced downwash at the tail.[24]

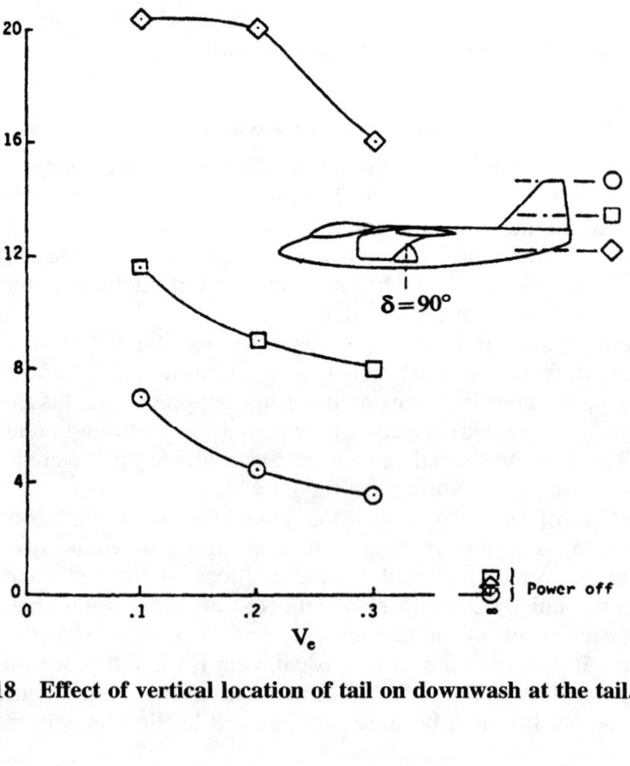

Fig. 3.18 Effect of vertical location of tail on downwash at the tail.[7]

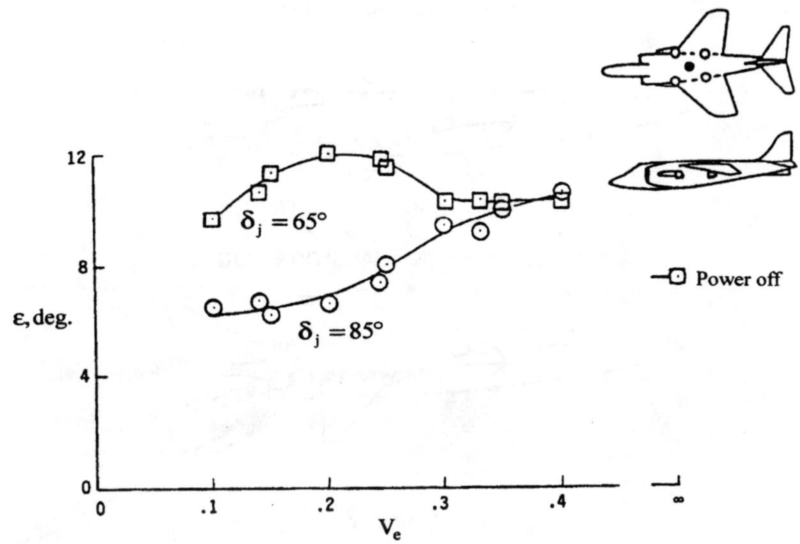

Fig. 3.19 Effect of flap deflection on the downwash experienced on a Harrier-type model.[6]

it also significantly alters the span load distribution. This change in span load distribution probably decreases the wing-lift-generated downwash faster than the downwash is increased, with increasing V_e, by the jet wake vortices.

Unlike the jet-induced moment on the wing body, which is generally independent of angle of attack, the downwash at the horizontal tail is a function of angle of attack and can, therefore, change both the trim and the stability of the tail-on configuration. The effect of power on the tail contribution to stability is highly dependent on the flowfield in which the tail operates and, in particular, on the flowfield generated inboard on the wing or by parts of the airplane ahead of the wing proper.

Figure 3.20 illustrates a change in flowfield at the tail between cruising and transition flight. On many modern high-speed jet fighters there are inlets or other such as fixed forewings of variable-sweep wings that produce and shed strong vortices inboard. It is desirable that the tail be located below or outboard of this trailing vortex system. Such arrangements would ensure that the tail would move away from the vortices as the angle of attack is increased. For jet VTOL airplane with an inboard tail, however, the situation is different. These inboard vortices can be pulled below the horizontal tail by the action of the lifting jets. Then as the angle of attack is increased, the tail is forced to traverse through these vortices. The severity of the problem this creates depends on many configuration variables, all of which have not been isolated.

A particularly severe example is presented in Fig. 3.21 for a four jet configuration with a fixed forewing and large inlets placed well forward.[25] With the power off, there is a linear and stable variation of pitching moment with angle of attack and a stable break at the stall. With power on, there is a

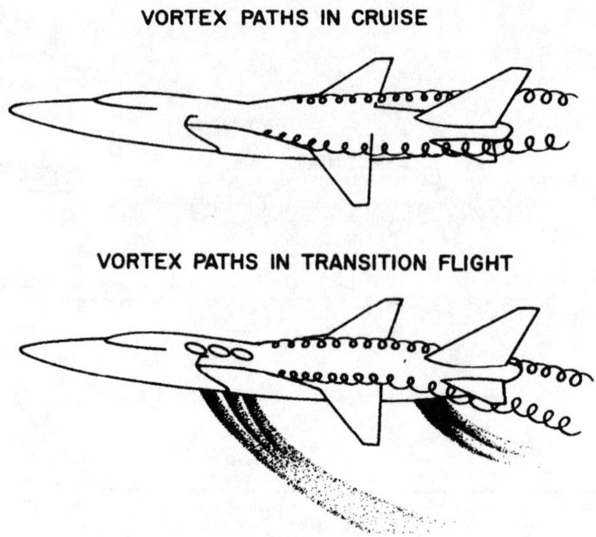

Fig. 3.20 Jet-induced effect on the flowfield at the tail for a configuration with strong lifting elements forward on the inboard part of the wing.

nose-up increment, but it is no longer invariant with angle of attack. It increases very rapidly as the angle of attack is increased and results in an extremely unstable configuration.

Data for two other configurations[26] are shown in Fig. 3.22. For the configuration with short inlets and no variable-sweep wing glove (on the left), the power effect on stability is essentially zero, as indicated by the fact that the two curves are nearly parallel. For the other configuration, which has long inlets and fixed forewing, a reduction in stability caused by power is encountered. However it

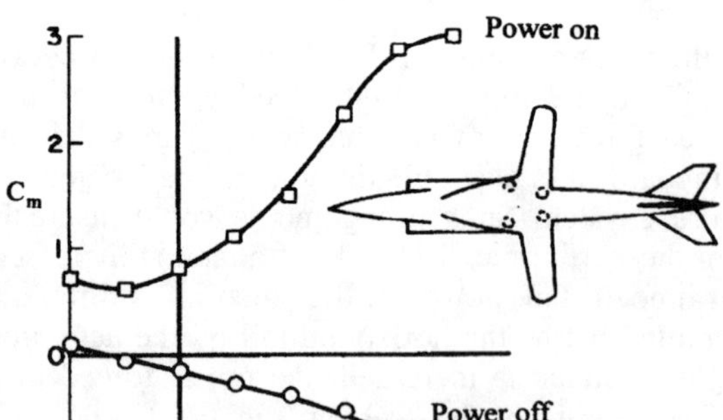

Fig. 3.21 Jet-induced effect on the longitudinal stability on a configuration with a large lifting forebody.[25]

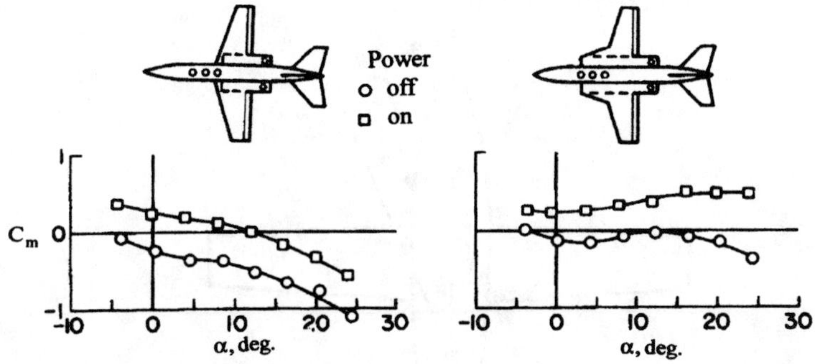

Fig. 3.22 Effect of configuration on the jet-induced effect on stability.[26]

is not as severe as that shown for the configuration in Fig. 3.21, which has a larger forebody, a small wing, and a longer tail length. The primary difference between the configurations appears to be in the size and length of the lifting elements forward of the wing, but the differences in tail length and configuration can also contribute.

Unfortunately these are only anecdotal data. There has not been a systematic investigation of the jet-induced effects on downwash at the tail, and on the effects on longitudinal stability, to provide a basis for a method to estimate these effects.

V. Lateral/Directional Characteristics

The same jet wake system that generates induced lift and pitching-moment increments can affect the rolling moment, yawing moment, and side-force characteristics of the configuration. As depicted schematically in Fig. 3.23, at an angle of sideslip β, the jet wake system is displaced laterally with respect to the configuration, and the pressure distribution that it generates on the body, and on the wing, is shifted toward the downstream side of the configuration thus generating a jet-induced rolling moment. An increment of side force and yawing moment is also induced on the body; although the jet wake system is usually far below the body, the force data available show that a significant sidewash is induced at the vertical tail. In addition the flow into the inlets will generate a side force (analogous to the inlet-momentum drag) and associated rolling and yawing moments.

There have been relatively few investigations of the jet exit flow on the lateral/directional characteristics experienced in the transition speed range. Reference 27 reviewed the data that were available (eight configurations at that time) and developed the relationships and rules of thumb presented in the following.

As with the longitudinal forces and moments, the total lateral/directional forces and moments can be expressed as the sum of the power-off forces and moments and the power-on jet-induced increments. For small angles:

Side force:

$$F_Y = C_{Y,\beta} q_o S_w \beta + \left(\frac{\Delta F_Y}{T}\right)_\beta T\beta \tag{3.49}$$

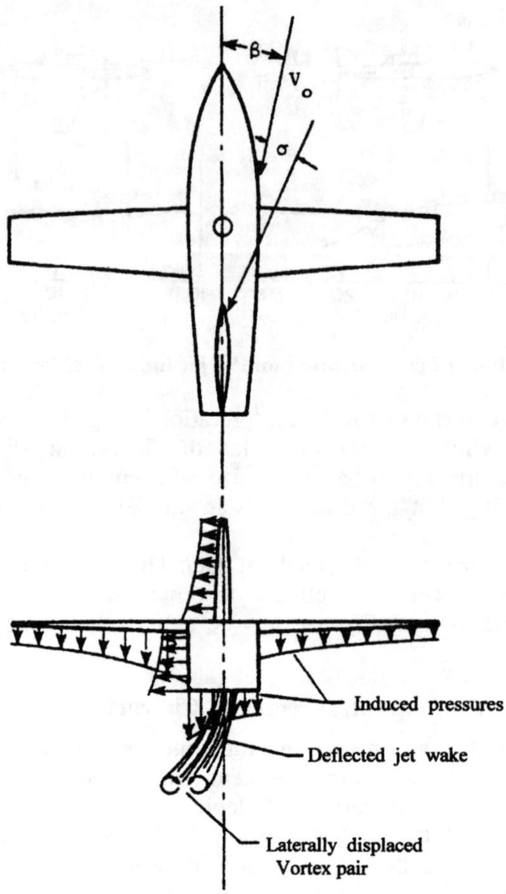

Fig. 3.23 Schematic of effect of sideslip on jet-induced pressures.

Yawing moment:

$$M_Z = C_{n,\beta} q_o S_w b \beta + \left(\frac{\Delta M_Z}{TD_e}\right)_\beta TD_e \beta \tag{3.50}$$

Rolling moment:

$$M_X = C_{\ell,\beta} q_o S_w b \beta + \left(\frac{\Delta M_X}{TD_e}\right)_\beta TD_e \beta \tag{3.51}$$

The induced increments ΔF_Y, ΔM_Z, and ΔM_X are produced by the pressures induced on the body, wing, and tail by the exit flows (Fig. 3.23) plus the inlet increments. The total induced effects are the sum of these induced increments and can be written as

$$\left(\frac{\Delta F_Y}{T}\right)_\beta = \left(\frac{\Delta F_Y}{T}\right)_{\beta,\text{inlet}} + \left(\frac{\Delta F}{T}\right)_{\beta,\text{body}} + \left(\frac{\Delta F_Y}{T}\right)_{\beta,\text{tail}} \tag{3.52}$$

$$\left(\frac{\Delta M_Z}{TD_e}\right)_\beta = \left(\frac{\Delta M_Z}{TD_e}\right)_{\beta,\text{inlet}} + \left(\frac{\Delta M_Z}{TD_e}\right)_{\beta,\text{body}} + \left(\frac{\Delta M_Z}{TD_e}\right)_{\beta,\text{tail}} \tag{3.53}$$

$$\left(\frac{\Delta M_X}{TD_e}\right)_\beta = \left(\frac{\Delta M_X}{TD_e}\right)_{\beta,\text{inlet}} + \left(\frac{\Delta M_X}{RD_e}\right)_{\beta,\text{body}} + \left(\frac{\Delta M_X}{TD_e}\right)_{\beta,\text{tail}} + \left(\frac{\Delta M_X}{TD_e}\right)_{\beta,\text{wing}} \tag{3.54}$$

The power-on and power-off stability derivatives $(M_X/TD_e)_\beta$, $(M_z/TD_e)_\beta$, and (F_Y/T_β) (analogous to the coefficient derivatives C_{l_β}, C_{n_β}, and C_{Y_β}) for one of the configurations (a model of the AV-6A Kestrel, Ref. 6) used in developing the method are shown in Fig. 3.24, and the increments that make up these estimates are presented in Fig. 3.25.

Although this configuration has a relatively low-aspect-ratio swept wing mounted high on the fuselage, the power-off dihedral effect (rolling moment

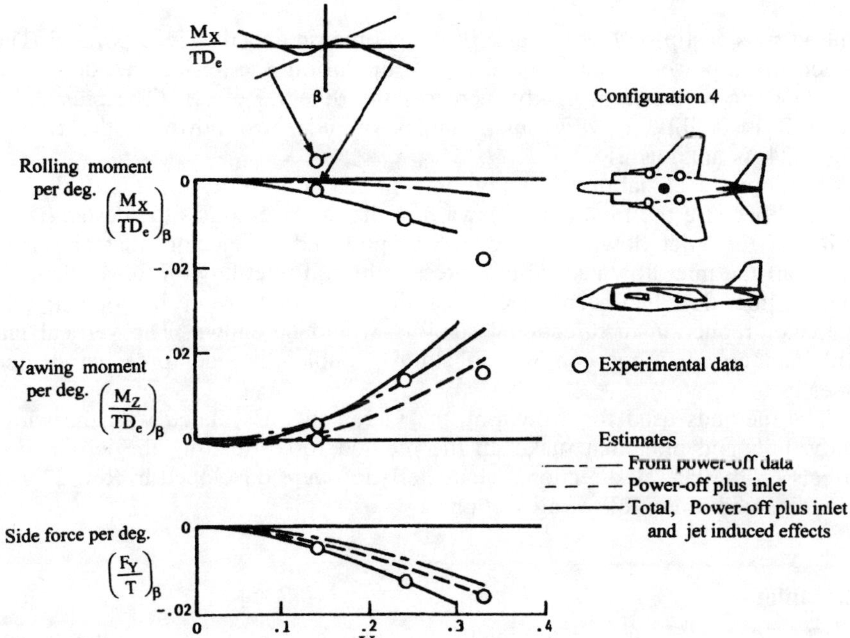

Fig. 3.24 Jet-induced effects on the lateral/directional characteristics of one configuration of Ref. 27.

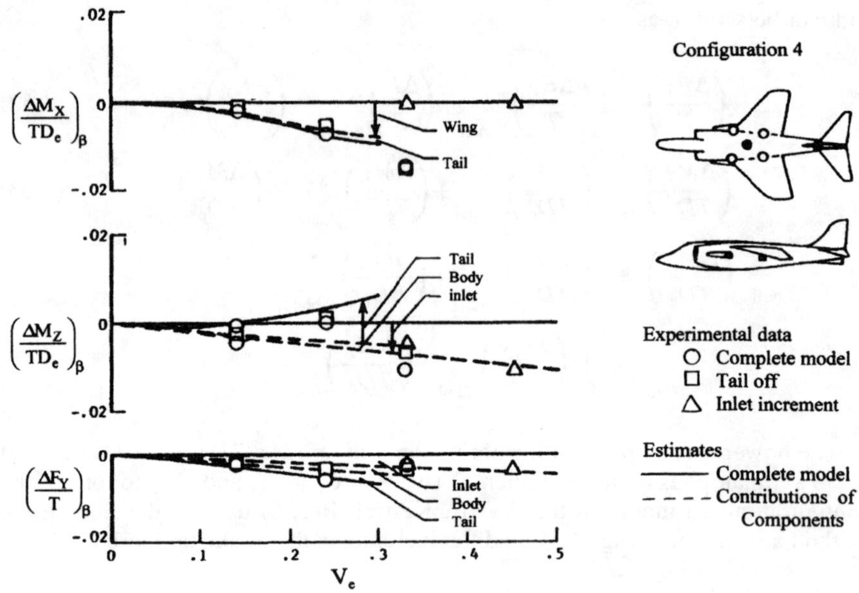

Fig. 3.25 Comparison of estimates with data.

caused by sideslip) is low because of the geometric anhedral incorporated. The effect of the power-on jet-induced effects in shifting the jet-induced download on the wing to leeward greatly increases the dihedral effect. (The reason for the roll instability at very small angles of sideslip, shown at the top of Fig. 3.24, is unknown.)

The directional stability (yawing moment caused by sideslip) is only slightly reduced because the favorable sidewash at the tail overcomes the destabilizing effect of the inlet flow. This model was powered by ejectors that entrained only half the inlet flow needed to represent the full inlet flow of the airplane. If the full inlet flow had been represented, the inlet effect would be doubled, and a greater reduction in directional stability would be shown. The vertical tail and inlet contributions combine to almost double the side force per degree sideslip.

The methods used for estimating the increments associated with the wing, body, tail, and inlet that make up the method for estimating the jet-induced effects on the lateral/directional characteristics were developed in Ref. 27 and are presented in the following sections.

A. Inlet

The side force caused by turning the flow into the inlets is the inlet-momentum force produced by the crossflow velocity $V_o \sin\beta$ and can be expressed as the inlet-momentum drag [Eq. (3.33)] multiplied by the sin of the sideslip angle.

For small angles the inlet side force per degree of sideslip is given by

$$\left(\frac{\Delta F_Y}{T}\right)_{\beta,\text{inlet}} = -0.0175\frac{w_i}{w_j}V_e \tag{3.55}$$

The yawing and rolling moments caused by turning the flow into the inlets is given by the side force multiplied by the effective arm. For yawing moment the arm is the longitudinal distance from the moment reference point to the center of the inlet face X_i and

$$\left(\frac{\Delta M_Z}{TD_e}\right)_{\beta,\text{inlet}} = \left(\frac{\Delta F_Y}{T}\right)_{\beta,\text{inlet}}\frac{X_i}{D_e} \tag{3.56}$$

The rolling-moment arm is the vertical distance to the effective center of action of the side-force increment caused by the inlet flow. For forward-facing inlets the arm is the vertical distance from the moment reference point to the center of the inlet face Z.

For upper-surface inlets the side force (like the drag in the preceding longitudinal mode) acts at a distance ΔZ [Eq. (3.48)] above the inlet face. The rolling moment caused by inlet flow is given by

$$\left(\frac{\Delta M_X}{TD_e}\right)_{\beta,\text{inlet}} = \left(\frac{\Delta F_Y}{T}\right)_{\beta,\text{inlet}}\frac{(Z_i + \Delta Z)}{D_e} \tag{3.57}$$

where ΔZ applies only in the case of upper-surface inlets.

For configurations with multiple inlets, the side-force, yawing-moment, and rolling-moment contributions of each must be calculated separately and summed.

B. Body

The pressures induced on the body by the jet, and its wake, give rise to a side force, yawing moment, and, to a lesser extent, rolling moment. The available data indicate that the variation of jet-induced side force with effective velocity ratio V_e is similar to that of the lift loss and that the magnitude of the side-force increment increases as the size of the configuration relative to the jet exit area increases. The jet-induced side force is estimated, for most configurations, by multiplying the lift loss $(\Delta L/T)_j$ estimated by Eq. (3.8) by a factor K_Y, which is a function of the ratio of the lateral profile area to jet area as shown in Fig. 3.26.

The body contribution to side force is given by

$$\left(\frac{\Delta F_Y}{T}\right)_{\beta,\text{body}} = \left(\frac{\Delta L}{T}\right)_{j,\text{body}} K_Y \tag{3.58}$$

where

$$K_Y = \left[0.001\frac{S_Y}{A_j} - 0.000004\left(\frac{S_Y}{A_j}\right)^2\right]K_{Y,n} \tag{3.59}$$

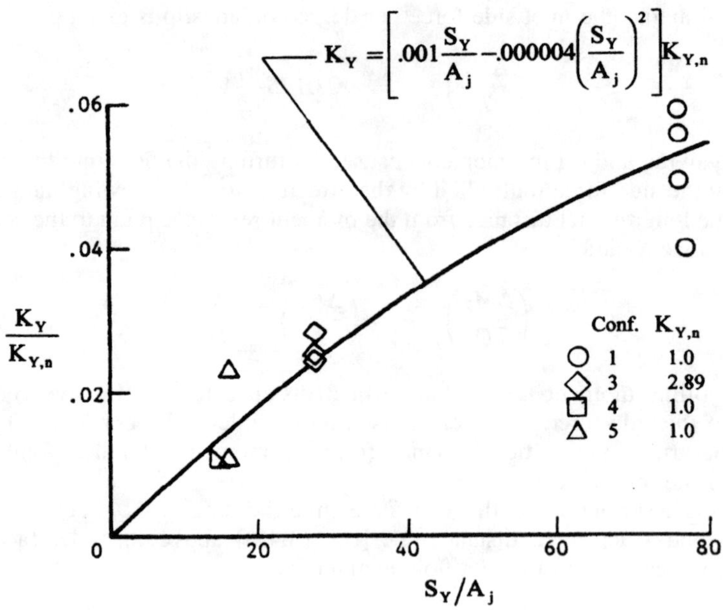

Fig. 3.26 Side-force factor K_Y (from Ref. 27).

and

$$K_{Y,n} = 1 + \left(\frac{\ell}{w} - 1\right)^{0.8} \tag{3.60}$$

The factor $K_{Y,n}$ is needed to account for the effect of slot nozzles or a row of closely spaced circular nozzles (see Fig. 3.13).

The yawing moment induced on the body, by a jet located at the moment reference point, is the side force multiplied by an effective arm:

$$\left(\frac{\Delta M_Z}{TD_e}\right)_{\beta,\text{body}} = \left(\frac{\Delta F_Y}{TD_e}\right)_{\beta,\text{body}} \left(\frac{X}{D_e}\right)_j \tag{3.61}$$

where the arm $(X/D_e)_j$ is the same arm used in estimating the jet-induced pitching moment [Eq. (3.35)].

The rolling moment induced on the body is caused by the lateral displacement of the pressure distribution induced on the lower surface of the body (lift loss) as indicated in Fig. 3.27.

For jets located within the body the rolling moment is the lift loss [Eq. (3.8)] multiplied by the same arm used in estimating the pitching moment [Eq. (3.35)] multiplied by the sine of the sideslip angle. For small angles the rolling moment is given by

$$\left(\frac{\Delta M_X}{TD_e}\right)_{\beta,\text{body}} = 0.0175\left(\frac{\Delta L}{T}\right)_{j,\text{body}} \left(\frac{X}{D_e}\right)_j \tag{3.62}$$

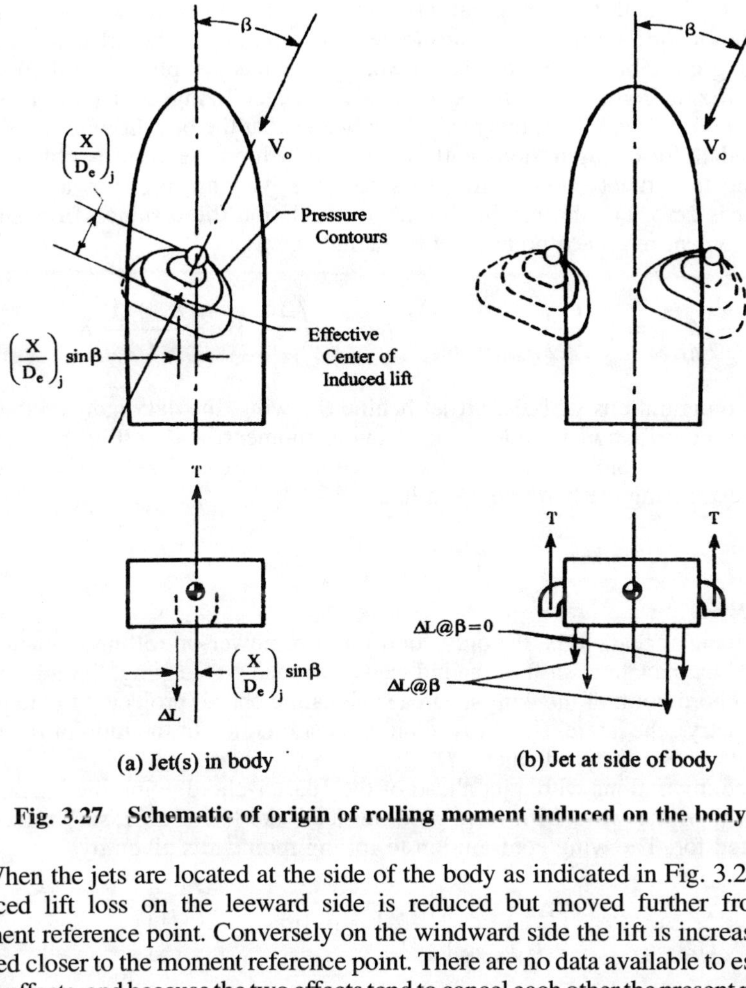

Fig. 3.27 Schematic of origin of rolling moment induced on the body.

When the jets are located at the side of the body as indicated in Fig. 3.27b, the induced lift loss on the leeward side is reduced but moved further from the moment reference point. Conversely on the windward side the lift is increased but moved closer to the moment reference point. There are no data available to estimate these effects, and because the two effects tend to cancel each other the present method assumes the rolling moment caused by jets on the side of the body is negligible.

With multiple-jet configurations that have a jet spacing far enough apart so that they cannot be treated as a slot jet (s/d greater than 1.0), the induced effects of each jet are estimated separately and summed.

$$\left(\frac{\Delta F_Y}{T}\right)_{\beta,\text{body}} = \left(\frac{\Delta L_f}{T}\right)_{j,\text{body}} K_{Y,f} + \left(\frac{\Delta L_r}{T}\right)_{j,\text{body}} K_{Y,r} \tag{3.63}$$

and

$$\left(\frac{\Delta M_X}{TD_e}\right)_{\beta,\text{body}} = 0.0175\left[\left(\frac{\Delta L_f}{T}\right)_{j,\text{body}}\left(\frac{X_f}{D_e}\right)_j + \left(\frac{\Delta L_r}{T}\right)_{j,\text{body}}\left(\frac{X_r}{D_e}\right)_j\right] \tag{3.64}$$

Most of the data for configurations for which the tail is off indicate that the wing-body contribution to yawing moment is stabilizing. The aft jets induce a stabilizing contribution. When the forward jet or jets are placed well forward, a destabilizing contribution that can exceed the stabilizing contribution of the aft jets is produced. In fact this analysis indicates that the best fit with the experimental data for configurations with the aft jets behind the wing is obtained by assuming the aft jets contribution to side force, yawing moment, and rolling moment is zero. For the yawing-moment calculation the distance from the jet to the moment reference point must be included.

$$\left(\frac{\Delta M_Z}{TD_e}\right)_{\beta,\text{body}} = \left(\frac{\Delta L_f}{T}\right)_{j,\text{body}} \frac{X_f + X'_{f'}}{D_e} K_{Y,f} + \left(\frac{\Delta L_r}{T}\right)_{j,\text{body}} \frac{X_r + X'_r}{D_e} K_{Y,r} \tag{3.65}$$

For configurations with the aft jet behind the wing, the data indicate that the rear jet's contribution to side force, yawing moment, and rolling moment is zero. For these configurations the second term in Eqs. (3.59–3.61) is set to zero in estimating the body contributions.

C. Wing

The wing appears to be the only contributor of power-on rolling moment and is caused by the lateral shift of the induced download on the wing. For a jet at the quarter-chord point of the wing mean aerodynamic chord (projected to the plane of symmetry), the lateral shift was found to be a function of the ratio of the wing span to the effective jet diameter D_e (Fig. 3.28).

For configurations with a jet ahead of the quarter-chord point, the lateral displacement of the wing with respect to the jet wake $[(X' - X_{c/4}) \sin \beta]$ must be accounted for. The wing contribution to rolling moment is given by

$$\left(\frac{\Delta M_x}{TD_e}\right)_{\beta,\text{wing}} = \left(\frac{\Delta L}{T}\right)_{j,\text{wing}} \left[\left(\frac{Y_w}{D_e}\right) + 0.0175 \frac{X' - X_{c/4}}{D_e}\right] K_{Y,n} \tag{3.66}$$

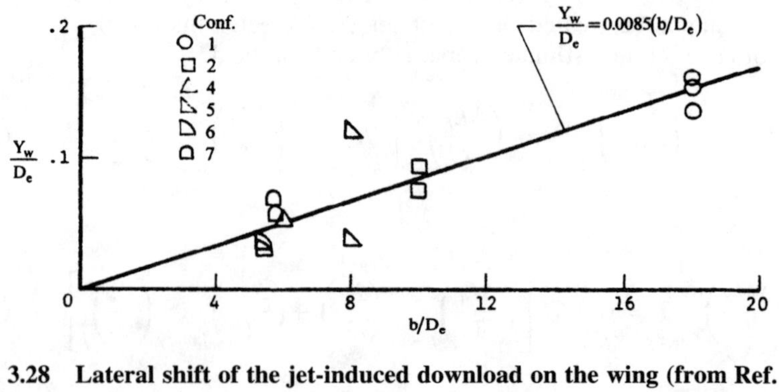

Fig. 3.28 Lateral shift of the jet-induced download on the wing (from Ref. 27).

where $K_{Y,n}$ [Eq. (3.60)] accounts for slot nozzles or a row of closely spaced jets (see Fig. 3.13) and

$$\frac{Y_w}{D_e} = \frac{0.0085b}{D_e}$$

For more widely spaced fore and aft jets the contribution of the front and rear jets must be estimated separately and summed.

$$\left(\frac{\Delta M_x}{TD_e}\right)_{\beta,\text{wing}} = \left(\frac{\Delta L_f}{T}\right)_{j,\text{wing}} \left[\left(\frac{Y_w}{D_e}\right) + 0.0175\frac{X_f' - X_{c/4}}{D_e}\right] K_{Y,n,f} + \left(\frac{\Delta L_r}{T}\right)_{j,\text{wing}} \left[\left(\frac{Y_w}{D_e}\right) + 0.0175\frac{X_r' - X_{c/4}}{D_e}\right] K_{Y,n,r} \tag{3.67}$$

Jets near, or aft of, the wing trailing edge induce a favorable lift. It has been found that these jets do not contribute to the jet induced rolling moment. For any configuration with jets near, or aft of, the wing trailing edge the second term in Eq. (3.67) is set to zero, and only the rolling moment from the first term is used.

D. Tail

The tail contributions to side force, yawing moment, and rolling moment are caused by the favorable sidewash induced at the vertical tail. This sidewash is produced primarily by the vorticity in the jet wake and would be expected to be inversely proportional to the distance of tail from the jet wake. However this distance is difficult to establish, and the sidewash has been correlated with the distance from the nozzle exit as shown in Fig. 3.29.

The side force induced on the vertical tail is given by

$$\left(\frac{\Delta F_Y}{T}\right)_{\beta,\text{tail}} = C_{Y,\beta,\text{tail}} \frac{\partial\sigma}{\partial\beta}\frac{S_w}{2A_j} V_e^2 \tag{3.68}$$

where $C_{Y,\beta,\text{tail}}$ is the power-off side-force coefficient slope (lift-curve slope) of the vertical tail. Power-off data can be used if available, or it can be calculated by the method of Ref. 22 (Sec. 9.1.2.1),

$$\frac{\partial\sigma}{\partial\beta} = 0.9\left[\sqrt{1.0 \Big/ \left(\frac{X_n}{D_e}\right)\left(\frac{Z_n}{D_e}\right)}\right] K_{Y,n} \tag{3.69}$$

and $K_{Y,n}$ [Eq. (3.60)] accounts for the effects of slot nozzles or a row of jets (Fig. 3.13).

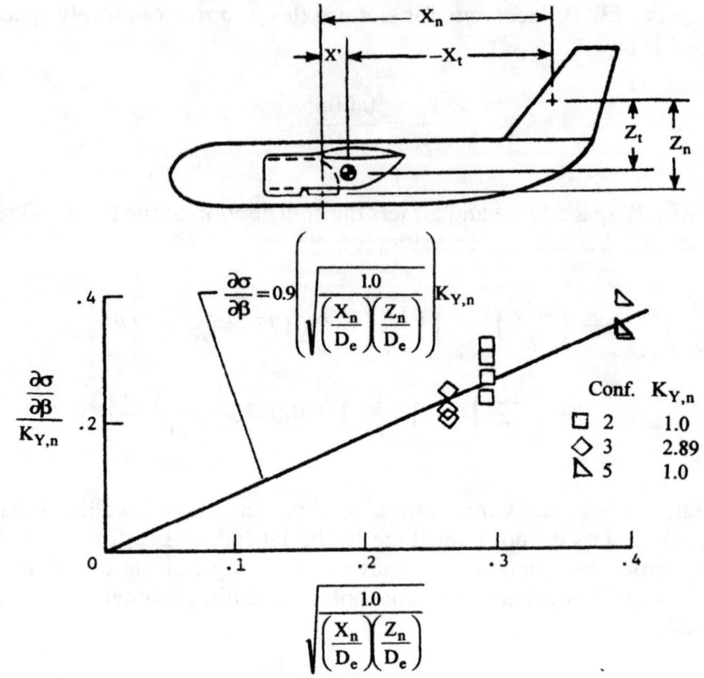

Fig. 3.29 Sidewash induced at the vertical tail by the jet (from Ref. 27).

For larger fore-and-aft spacing the sidewash caused by the front and rear jets must be calculated separately and summed. The sidewash is given by

$$\frac{\partial\sigma}{\partial\beta} = 0.9\left[\sqrt{1.0\Big/\left(\frac{X_{n,f}}{D_e}\right)\left(\frac{Z_{n,f}}{D_e}\right)}\right]K_{Y,n,f} + 0.9\frac{V_{e,r}}{V_{e,f}}\left[\sqrt{1.0\Big/\left(\frac{X_{n,r}}{D_e}\right)\left(\frac{Z_{n,r}}{D_e}\right)}\right]K_{Y,n,r} \tag{3.70}$$

where the effective velocity ratio $V_{e,r}$ at the rear jet is reduced because it is operating in the wake of the front jet and is calculated by Eq. (3.22).

For configurations with the aft jet behind the wing, only the sidewash induced by the front jet is used.

The yawing-moment increment caused by the vertical tail is given by

$$\left(\frac{\Delta M_Z}{TD_e}\right)_{\beta,\text{tail}} = \left(\frac{\Delta F_Y}{T}\right)_{\beta,\text{tail}}\frac{X_t}{D_e} \tag{3.71}$$

The rolling-moment increment caused by the vertical tail is given by

$$\left(\frac{\Delta M_X}{TD_e}\right)_{\beta,\text{tail}} = \left(\frac{\Delta F_Y}{T}\right)_{\beta,\text{tail}} \frac{Z_t}{D_e} \tag{3.72}$$

VI. Reaction Controls

In hover and at low speeds in transition flight (where conventional aerodynamic control surfaces are ineffective), jet V/STOL aircraft often use reaction jets to provide or augment control. Typically air is bled from the engine compressor and ducted to nozzles at the extremities of the airframe to provide the control moments needed.

In hover the moment produced by the roll control jets is simply the thrust of the reaction control jet multiplied by the distance from the moment reference point to the jet.

$$\Delta M_{rj@V_e=0} = T_{rj}\left(\frac{b}{2} - y\right) \tag{3.73}$$

where y is the distance the control jet is inboard of the wing tip.

At forward speed the interaction with the freestream causes the control jet to be deflected aft and rolled up into a pair of vortices. As discussed earlier (Secs. I and II) these vortices, along with the blockage and entrainment action of the jet, induce suction pressures beside and behind the jet (Fig. 3.30). The integrated effect of these pressures is an effective reduction in the net thrust of the control jet and the moment produced.

The effect of the freestream on the effectiveness of roll control jets located near the wing tip was investigated in Ref. 28. This study showed that the loss in rolling moment produced by such jets at transition speeds depended on the proximity of the control jet to the wing tip and wing trailing edge (Fig. 3.31).

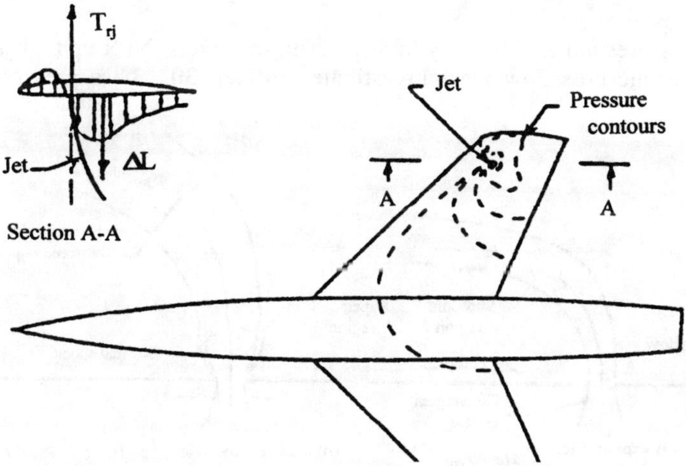

Fig. 3.30 Schematic of control-jet/freestream interaction.

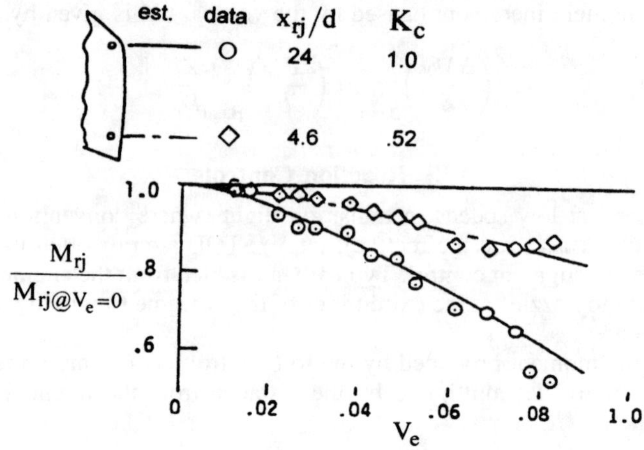

Fig. 3.31 Effect of control jet position on jet-induced rolling moment.

The method presented here for estimating the effects of control jet/freestream interaction on the effectiveness of roll control jets was developed in appendix H of Ref. 29.

Because compressor bleed air is often used for the reaction control jets, the jet pressure ratios (the NPR can be of the order of 20) involved are much higher than those from the primary lifting nozzles (NPR of the order of 4 to 6). With a fully expanded jet (subcritical pressure ratio jet issuing from a convergent nozzle) the deflection begins close to the surface of the wing. With an underexpanded jet (supercritical pressure ratio jet issuing from a convergent nozzle) there is a supersonic region outside the nozzle (Fig. 3.32) that resists deflection. Also the diameter of this supersonic region is increased. As the pressure ratio increases and these changes become more pronounced, the jet-induced suction pressures increase.

The pressures induced by very high-pressure-ratio jets (NPR up to 45) issuing into a subsonic crossflow were investigated in Ref. 30. These pressures were

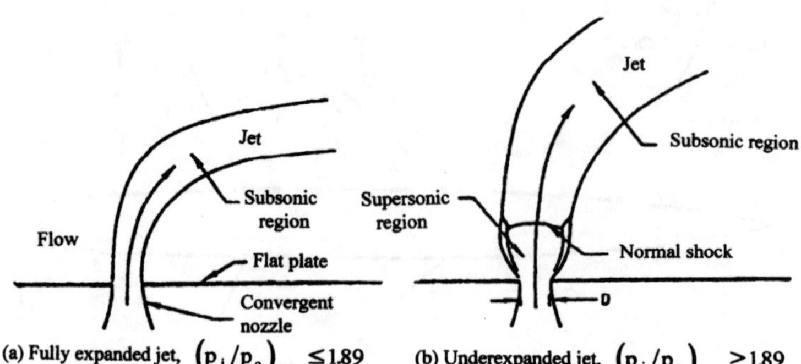

Fig. 3.32 Schematics of a subcritical jet and a supercritical jet issuing into a subsonic crossflow.

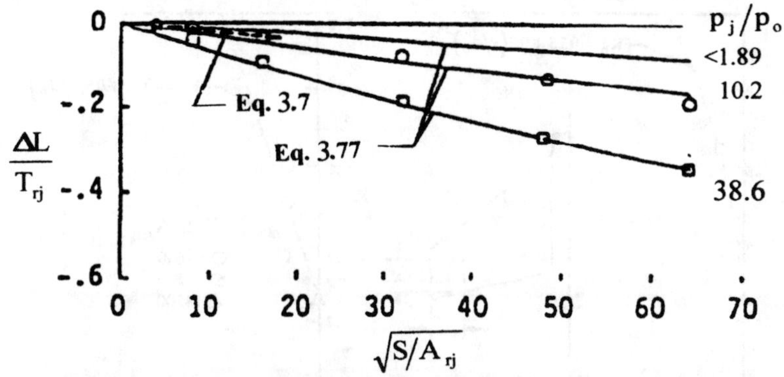

Fig. 3.33 Lift loss for supercritical pressure ratios.

integrated in Ref. 30 to obtain the lift losses. These lift losses are compared with those for the main jets at subcritical pressure ratios [Eq. (3.7)] in Fig. 3.33.

For the present method the lift loss as a result of jet-induced effects [Fig. 3.33 and shown later as Eq. (3.77)] is assumed to be made up of two terms, one for subcritical jets pressure ratios and an additional increment for supercritical pressure ratio:

$$\frac{\Delta L}{T_{rj}} = \left[\left(\frac{\Delta L}{T}\right)_o + \left(\frac{\Delta L}{T}\right)_p\right] K_b K_c \tag{3.74}$$

where

$$\left(\frac{\Delta L}{T}\right)_o = (3V_e^3 - 2.4V_e^2)\left(\frac{S}{A_{rj}}\right)^{0.5} + 0.41V_e^{2.2}\left(\frac{S}{A_{rj}}\right)^{0.688} \tag{3.75}$$

For $p_j/p_o < 1.89$

$$\left(\frac{\Delta L}{T}\right)_p = 0 \tag{3.76}$$

and for $p_j/p_o > 1.89$

$$\left(\frac{\Delta L}{T}\right)_p = -0.017V_e\left(\frac{S}{A_{rj}}\right)^{0.42}\left(\frac{p_j}{p_o} - 1.89\right)^{0.75} \tag{3.77}$$

The factors K_b and K_c account for the location of the jets with respect to the wing tip and wing trailing edge, respectively. These factors (Fig. 3.34) were derived from the data of Ref. 28 and unpublished data on a delta-wing configuration. These data show that when the jet is located approximately 10 or more diameters inboard of the wing tip and 12 or more diameters ahead of the wing

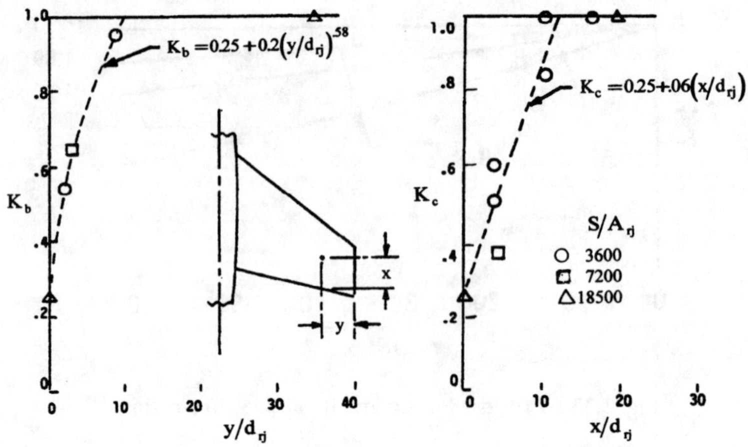

Fig. 3.34 Effect of control jet location with respect to the wing tip and to the wing trailing edge.

trailing edge the values can be taken as

$$K_b = K_c = 1.0 \tag{3.78}$$

When the control jet is closer to the wing tip or trailing edge,

$$K_b = 0.25 + 0.2\left(\frac{y}{d_{rj}}\right)^{0.58} \tag{3.79}$$

and

$$K_c = 0.25 + 0.06\left(\frac{x}{d_{rj}}\right) \tag{3.80}$$

The roll control moment available in the transition speed range is the control jet thrust minus the induced lift loss multiplied by the moment arm:

$$M_{rj} = T_{rj}\left(1.0 - \frac{\Delta L}{T_{rj}}\right)\left(\frac{b}{2} - y\right) \tag{3.81}$$

and the ratio of the moment available in transition to that produced in hover is

$$\frac{M_{rj}}{M_{rj@V_e=0}} = 1.0 - \left[\left(\frac{\Delta L}{T}\right)_o + \left(\frac{\Delta L}{T}\right)_p\right] K_b K_c \tag{3.82}$$

A. Effect of Sideslip

A limited investigation of the effects of sideslip on control jet effectiveness was included in the investigation of Ref. 28. Data were taken only at a sideslip

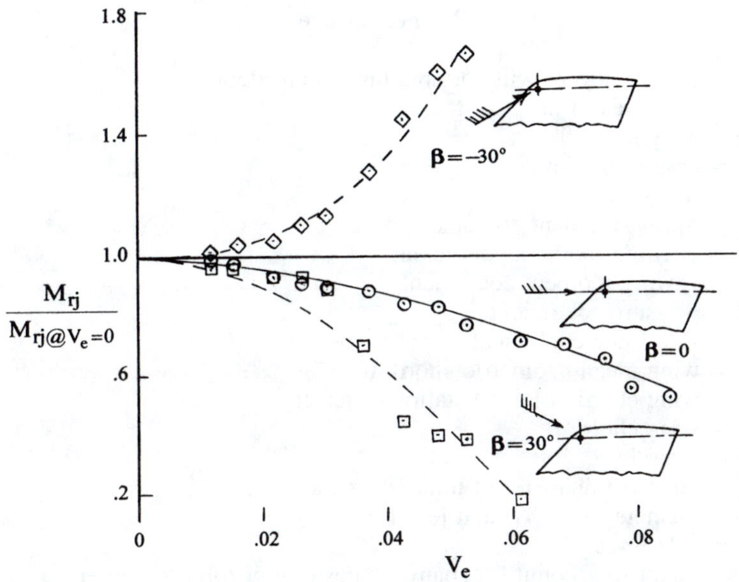

Fig. 3.35 Effect of sideslip on roll control jet effectiveness.

angle of 30 deg (Fig. 3.35). As would be expected, for a downward-directed jet on the windward wing ($\beta = +30$ deg) the suckdown is increased because more of the pressure pattern induced by the jet falls on the wing. The very large favorable effect on the leeward wing ($\beta = -30$ deg) at first seemed surprising and was discounted in Ref. 28 because it would be expected that more of the induced negative pressure region would fall "off" the wing and the induced effects should be reduced. However the favorable effect shown here is probably related to the jet flap lift gain presented in Sec. IV (see Fig. 3.12). That is, to the crossflow component of velocity the wing tip acts like a wing trailing edge, and favor pressures are induced on both the lower and upper surfaces.

These effects of sideslip are favorable for a downward-directed control jet. In a sideslip condition a large rolling moment is induced by the induced effects of the main lifting jets being shifted to the leeward wing. A downward-directed jet on the leeward wing would experience the large favorable effect of sideslip shown in Fig. 3.34. Adverse effects would be encountered only if an upward-directed jet on the windward wing were used to attempt to counter the rolling moment caused by the main lifting jets.

B. Effect of Ailerons

The effect of aileron deflection on the jet-induced rolling moments was also investigated in Ref. 28. These effects were found to be negligible.

Nomenclature

A = aspect ratio of wing or area under consideration
A_j = total area of all jets, ft^2
b = wing span, ft
C_D = drag coefficient
C_L = lift coefficient
C_ℓ = rolling-moment coefficient
C_m = pitching-moment coefficient
C_n = yawing-moment coefficient
C_p = pressure coefficient
C_Y = side-force coefficient
$\bar{c}$ = wing mean geometric chord, ft
c_j = wing chord at lateral station of jet, ft
c = wing chord, ft
D = drag, lb
D_e = effective diameter of total jet exit area, ft
d = diameter of individual jet, ft
F_Y = side force, lb
K_b = factor to account for spanwise position of roll control jet [Eq. (3.79)]
K_c = factor to account for chordwise position of roll control jet [Eq. (3.80)]
$K_{L,A}$ = factor for correcting lift for the effects of planform aspect ratio [Eqs. (3.9) and (3.10)]
$K_{\mathrm{L,flap}}$ = factor that accounts for presence or absence of flap on induced lift [Eqs. (3.31) and (3.32)]
$K_{L,n}$ = factor for correcting lift for the effects of width/length ratio of jets [Eq. (3.21)]
$K_{L,x}$ = factor for correcting lift for the effects of jet longitudinal position [Eqs. (3.11–3.13)] for jet loss and [Eqs. (3.27–3.29)] for "jet flap" lift gain
$K_{L,y}$ = factor for correcting lift for the effects of lateral spacing of side-by-side jets [Eqs. (3.15) and (3.16)]
$K_{L,y'}$ = factor for correcting lift for the effects of lateral distance from side of body [Eqs. (3.17–3.19)]
$K_{L,z}$ = factor for correcting lift for the effects of wing height [Eq. (3.14)] for jet loss and [Eq. (3.30)] for "jet flap" lift gain
$K_{m,A}$ = factor for correcting pitching moment for effects of planform aspect ratio [Eq. (3.37)]
$K_{m,n}$ = factor for correcting pitching moment for effects of length/width ratio of jets [Eq. (3.41)]
$K_{m,x}$ = factor for correcting pitching moment for effects of longitudinal position of jet [Eqs. (3.39) and (3.40)]
$K_{m,y}$ = factor for correcting pitching moment for effects of lateral spacing of side-by-side jets [Eqs. (3.42) and (3.43)]
K_Y = factor used in estimating jet-induced side force [Eq. (3.59)]
$K_{Y,n}$ = factor for correcting side force for effects of length/width ratio of jets [Eq. (3.60)]
K_1 = factor used in calculating $K_{m,A}$ [Eq. (3.38)]
L = lift, lb

ℓ = length of slot jet or row of circular jets (see Fig. 3.13), ft
M = pitching moment, ft · lb
M_{rj} = rolling moment caused by roll control jet, ft · lb
M_X = rolling moment, ft · lb
M_Z = yawing moment, ft · lb
N = number of jets
q_j = jet dynamic pressure, lb/ft^2
q_0 = freestream dynamic pressure, lb/ft^2
S = total planform area, or area under consideration, ft^2
S_f = area forward of the jet, ft^2
S_Y = lateral profile area of body, ft^2
s = longitudinal space between jets (see Fig. 3.13), ft
T = total jet thrust, or thrust of jets under consideration, lb
T_{rj} = thrust of roll control jet, lb
V_e = effective velocity ratio, $\equiv \sqrt{q_0/q_j}$
V_0, V_∞ = freestream velocity, ft/s
w = width of slot jet or row of circular jets (see Fig. 3.13), ft
w_i = weight flow of air entering the inlet, lb/s
w_j = weight flow of the jets stream, lb/s
X = effective arm at which induced lift acts, ft
$X_{c/4}$ = longitudinal distance of quarter-chord point of mean geometric chord from moment reference point; positive when quarter-chord point is ahead of moment reference point, ft
X_i = longitudinal distance from moment reference point to center of inlet face, ft
X_j = longitudinal distance from moment reference point to quarter-chord point of exposed wing; positive for quarter-chord point ahead of moment reference point, ft
X_n = longitudinal distance from nozzle center to aerodynamic center of vertical tail (see Fig. 3.29), ft
X_{rj} = longitudinal distance roll control jet is aft of leading edge, ft
X_Γ = longitudinal distance from moment reference point to midchord point of exposed wing; positive for midchord point ahead of moment reference point, ft
X_t = longitudinal distance from moment reference point to aerodynamic center of vertical tail (see Fig. 3.29), ft
X' = longitudinal distance between a jet and the moment reference point; positive when jet is ahead of moment reference point, ft
x_j = distance from why leading edge to lateral station of jet, ft
$Y_{w'}$ = lateral shift of induced download on wing, ft
y = lateral distance between jet centers (see Fig. 3.13), ft
y' = lateral distance of jet from side of body (see Fig. 3.13), ft
Z_i = effective arm at which inlet force acts, ft
Z_n = vertical distance from nozzle exit plane to aerodynamic center of vertical tail (see Fig. 3.29), ft
Z_t = vertical distance from moment reference point to aerodynamic center of vertical tail (see Fig. 3.29), ft
z = height of wing above jet exit plane, ft

ΔD = increment of drag caused by flow into inlet, lb
ΔL = jet-induced lift increment, lb
ΔM = jet-induced increment of moment, ft · lb
α = angle of attack, deg
β = sideslip angle, deg
ΔZ = increment of vertical distance to effective center of action of inlet force for upper surface inlets [Eq. (3.48)], ft
δ_f = flap deflection, deg
δ_j = jet deflection angle, measured from the horizontal, deg
ε = downwash angle at the tail, deg
σ = sidewash angle, deg
ϕ = lateral splay angle of jets, deg

Subscripts

b = body
basic = basic jet induced increment
f = forward of the jet, or front jet
i = inlet
j = jet induced
t = transition
r = rear jet
rj = roll control jet
α = angle of attack
β = sideslip, per degree
Γ = jet flap effect
δ = jet deflection angle
ϕ = splay angle
∞ = hover out of ground effect

References

[1]Margason, R. J., "Fifty Years of Jet in Cross Flow Research." AGARD, CP 534 April 1993.

[2]Williams, J., and Wood, M. N., "Aerodynamic Interference Effects with Jet-Lift V/STOL Aircraft Under Static and Forward-Speed Condition," Royal Aircraft Establishment, Technical Rept. 66403, Farnborough, England, Dec. 1966.

[3]Carter, A. W., "Effects of Jet-Exhaust Location on the Longitudinal Aerodynamic Characteristics of a Jet V/STOL Model," NASA TN D-5333, July 1969.

[4]"Symposium on Analysis of a Jet in a Subsonic Crosswind," NASA SP-218, Sept. 1969.

[5]Wooler, P. T., Kao, H. C., Schwendemann, M. F., Wasson, H. R., and Ziegler, H., "V/STOL Aircraft Aerodynamic Prediction Methods Investigation," Air Force Flight Dynamics Lab., AFFDL-TR-72-26, Wright Patterson Air Force Base, Ohio, Jan. 1972.

[6]Margason, R. J., Vogler, R. D., and Winston, M. M., "Wind-Tunnel Investigation at Low Speeds of a Model of the Kestrel (XV-6A) Vectored-Thrust V/STOL Airplane," NASA TN D-6826, July 1972.

[7]Mineck, R. E., and Schwendemann, M. F., "Aerodynamic Characteristics of a Vectored-Thrust V/STOL Fighter in the Transition-Speed Range," NASA TN D-7191, May 1973.

[8]Fearn, R. L., and Weston, R. P., "Velocity Field of a Round Jet in a Crossflow for Various Jet Injection Angles and Velocity Ratios," NASA Technical Paper 1506, Oct. 1979.

[9]Beatty, T. D., and Kress, S. S., "Prediction Methodology for Propulsive Induced Forces and Moments of V/STOL Aircraft in Transition/STOL Flight," NAVAIRDEVCEN Rept. NADC-77119-30, Naval Air Development Center, Warminster, PA, July 1979.

[10]Perkins, S. C., Jr., and Mendenhall, M. R., "A Study of Real Jet Effects on the Surface Pressure Distribution Induced by a Jet in a Crossflow," NASA CR-166150, March 1981.

[11]Aoyagi, K., and Snyder, P. K., "Experimental Investigation of a Jet Inclined to a Subsonic Crossflow," AIAA/NASA Ames V/STOL Conference, AIAA Paper 81-2610, Dec. 1981.

[12]Kotansky, D. R., "Jet Flow Fields," *Special Course on V/STOL Aerodynamics*, Advisory Group for Aerospace R & D AGARD, Rept. 710, France, 1984, pp. 7-1–7-48.

[13]Roth, K. R., "Numerical Simulation of a Subsonic Jet in a Crossflow," *Proceedings of the International Powered Lift Conference*, Society of Automotive Engineers SAE P-203, Dec. 1987, pp. 425–432.

[14]McMahon, H. M., and Mosher, D. K., "Experimental Investigation of Pressures Induced on a Flat Plate by a Jet Issuing Into a Subsonic Crosswind," *Symposium on Analysis of a Jet in a Subsonic Crosswind*, NASA Langley Research Center NASA SP-218, 1969, pp. 49–62.

[15]Margason, R. J., "Propulsion-Induced Effect Caused by out-of-Ground Effects," SAE 872307, Also in *Proceedings of the International Powered Lift Conference* SAE P-203, 1987, pp. 31–57.

[16]Hunt, D. L., and Bruce, R. J., "The Role of CFD in Predicting Vectored Jet Interference for STOVL Aircraft," *International Powered Lift Conference*, London, U.K., 1998, pp. 11.1–11.12.

[17]Johnson, D. B., Lacey, T. R., and Voda, J. J., "Powered Wind Tunnel Testing of the AV-8B: A Straight Forward Approach Pays off," 17th Aerospace Sciences Meeting AIAA Paper 79-0333, New Orleans, LA, Jan. 1979.

[18]Margason, R. J., "Review of Propulsion-Induced Effects on Aerodynamics of Jet/STOL Aircraft," NASA TN D-5617, Feb. 1970.

[19]Saddington, A. J., and Knowles, K., "A Review of out-of-Ground-Effect Propulsion Induced Interference on STOVL Aircraft," *Progress in Aerospace Sciences*, Vol. 41, June 2005, pp. 175–191.

[20]Margason, R., "Correlation of STOVL Jet Induced Effects with Examples For Configurations with Supersonic Convergent Jets," AIAA Paper 2002-5867, Nov. 2002.

[21]Henderson, C., Clark, J., and Walters, M., "V/STOL Aerodynamics and Stability and Control Manual," NAVAIRDEVCEN, Rept. NADC80017-60, Naval Air Development Center, Warminster, PA, Jan. 1980.

[22]Stewart, V. R., and Kuhn, R. E., "A Method for Predicting the Aerodynamic Stability and Control Parameters of STOL Aircraft Configurations," AFWAL, TR-87-3019, Vol. II, Wright Patterson Air Force Base, Ohio, June 1987.

[23]McCormick, B. W., Jr., *Aerodynamics of V/STOL Flight*, Academic Press, New York, 1967, pp. 194–211.

[24]Margason, R. J., "Jet-Induced Effects in Transition Flight," NASA SP-116, April 1966, pp. 177–189.

[25]Vogler, R. D., and Kuhn, R. E., "Longitudinal and Lateral Stability Characteristics of Two Four-Jet VTOL Models in the Transition Speed Range," NASA TM X-1092, 1965.

[26]Margason, R. J., and Gentry, G. L., Jr., "Aerodynamic Characteristics of a Five-Jet VTOL Configuration in the Transition Speed Range," NASA TN D-4812, Oct. 1968.

[27]Kuhn, R. E., "An Engineering Method for Estimating the Lateral/Directional Characteristics of V/STOL Configurations in Transition," NAVAIRDEVCEN, Rept. NADC-81031-60, Naval Air Development Center, Warminster, PA, Feb. 1981.

[28]Spreeman, K. P., "Free Stream Interference Effects on the Effectiveness of Control Jets near the Wing Tip of a VTOL Aircraft Model," NASA TN D-4084, Aug. 1967.

[29]Stewart, V. R., and Kuhn, R. E., "A Method for Predicting the Aerodynamic Stability and Control Parameters of STOL Aircraft Configurations," AFWAL, TR-87-3019, Vol. III (Appendix H), Wright Patterson Air Force Base, Ohio, June 1987.

[30]Shaw, C. S., and Margason, R. J., "An Experimental Investigation of a Highly Underexpanded Sonic Jet Ejecting from a Flat Plate into a Subsonic Crossflow," NASA TN D-7314, Dec. 1973.

Chapter 4

STOL Operation (Transition-in-Ground Effect)

WHEN THE aircraft is operating close to the ground at transition speeds, that is, in STOL operation, all of the flow phenomena just discussed are present but modified. The suckdown and fountain effects induced in hover are modified by the crossflow, and the lift loss and pitching moments induced by the deflected jet and its wake are modified by the proximity to the ground.

In addition a ground vortex (Fig. 4.1) is formed by the action of the freestream in opposing the wall jet flowing forward from the impingement point of the front jet (Refs. 1–11). This ground vortex induces additional suckdown pressures, primarily ahead of the jet(s). The ground vortex is also one of the primary mechanisms contributing to hot-gas ingestion, discussed in a later section.

Early methods for estimating the lift and moment induced by the ground vortex (Refs. 8 and 9) were based primarily on force data. The method presented here is based on that presented in Ref. 10, which was based on detailed pressure distribution data that provided a more detailed understanding of the origin of the forces and moments generated. Another experimental investigation of the ground vortex flow is presented in Ref. 11, where quantitative measurements using laser-doppler-anemometer (LDA) and laser-induced-fluorescence techniques (LIF) were presented.

I. Ground Vortex

As shown by the typical centerline distributions presented in Fig. 4.2, the ground vortex induces negative, or suction, pressures on the lower surface of the configuration as well as on the ground, in the region of the vortex, and positive pressures ahead of the ground vortex.

A. Zero Pressure Line—Single Jet

The detailed pressure distributions of Ref. 10 provide information on the shape and position of the ground vortex generated by a single jet. Typical pressure distributions (at one crossflow velocity) at various spanwise stations are presented in Fig. 4.3. The peak negative pressure induced in the region of the vortex decreases with spanwise location, also the point at which the pressure changes form positive ahead of the vortex to negative in the region of the vortex, moves aft with spanwise location. The line connecting these points is referred to as the zero pressure line, and its location and shape are critical elements of the method for estimating the lift and moment induced by the ground vortex. The

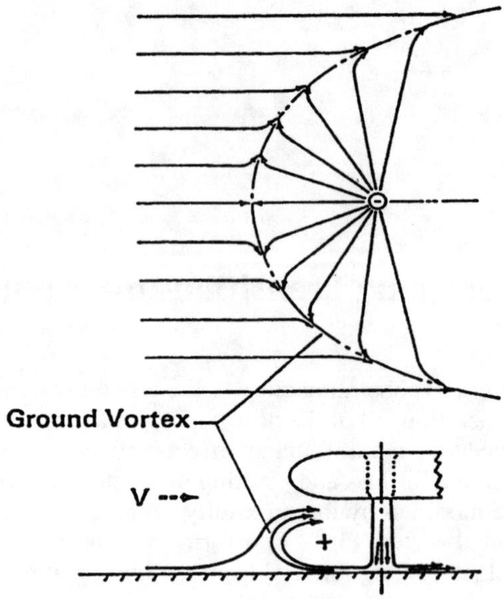

Fig. 4.1 Formation of the ground vortex.

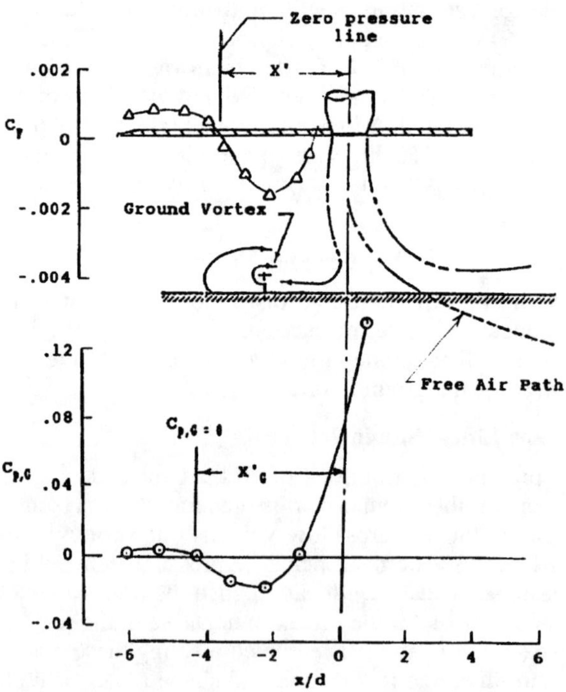

Fig. 4.2 Centerline pressure distribution on the lower surface (top) and on the ground (bottom).

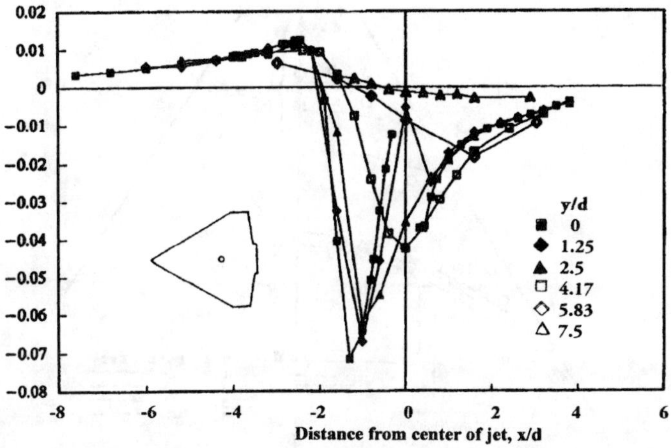

Fig. 4.3 Chordwise pressure distributions at several spanwise stations; $h/d = 1.7$, and $V_e = 0.2$.

line is parabolic in shape, as shown in Fig. 4.4, and for a single jet, the distance X' (on the centerline) from the jet to the zero pressure line, is given by

$$\frac{X'}{d} = 0.6\,K_{\text{gb}}\left(\frac{S}{A_j}\right)^{0.2} V_e^{-0.4}\left(\frac{h}{d}\right)^{0.06V_e^{-0.7}} + \frac{h}{d}\tan(\delta - 90) \tag{4.1}$$

where K_{gb} is the factor that adjusts for the ground situation

$K_{\text{gb}} = 0.67$ for an aircraft moving over the ground, or a model over a moving belt ground board

$K_{\text{gb}} = 1.0$ for an aircraft hovering in a head wind, or a model over a fixed ground board

and where d is the jet diameter, h is the lower surface height above the ground, V_e is the effective velocity ratio, and δ is the deflection of the jet measured aft from the vertical.

The shape of the zero pressure line caused by the ground vortex generated by a single jet, or the front jet of a tandem jet arrangement (Fig. 4.4) is given by

$$y = 2\sqrt{X'(X' - x)} \tag{4.2}$$

B. Zero Pressure Line—Side-by-Side Pair

The preceding applies only for a single circular jet. A comparable method is not available for estimating the location of the zero pressure line for noncircular jets or a side-by-side pair of jets. However the data of Ref. 12 show that the

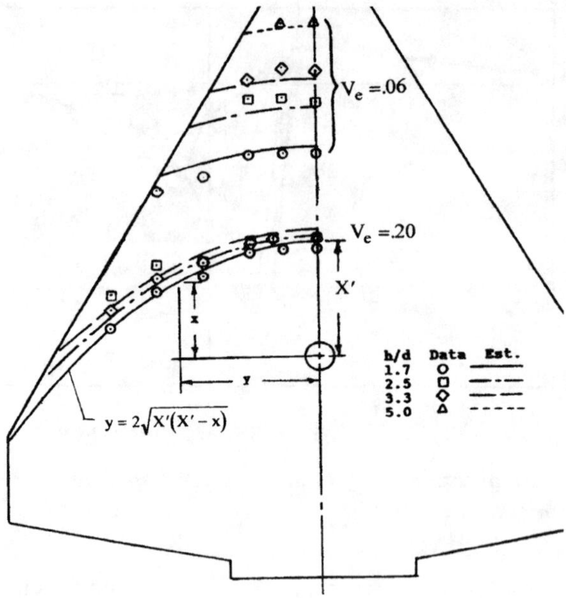

Fig. 4.4 Location of the zero pressure line on the lower surface of the configuration.

forward penetration of the zero pressure line produced by a side-by-side pair of circular jets is less than that for the equivalent single circular jet (Fig. 4.5).

Although there is scatter in both the single jet and jet pair data, the data for the pair are even further aft than the estimates would indicate. A similar finding is reported in Ref. 13, where the zero pressure point (on the ground, rather than on the configuration lower surface as in Ref. 12) was also found to be further aft. Apparently the part of the entrainment of the fountain flow between the pair of jets is achieved by entraining air from the upper surface of the forward-flowing wall jets, thus reducing their energy and ability of the wall jet to flow forward against the freestream.

This is in marked contrast to the observed location of the zero pressure line of the ground vortex generated by a pair of jets as shown in Fig. 4.6 (Ref. 5). With side-by-side jets (3 d spacing) the wall jets flowing outward from the impingement points meet and produce a fan-shaped fountain flow similar to that depicted in Fig. 1.8. Penetration is increased with twin nozzles over that expected for a single jet of the same total exit area and the same V_e. However this increased penetration is confined to the central upwind portion of the vortex where the twin jet fountain projects forward of the vortex produced by a single jet. Away from this region the twin and single nozzle flowfields are virtually identical. It was noted in Ref. 5 that this result might be dependent on the lateral spacing of the twin jets. Wider spacing did not produce the spike in the locus on the centerline.

On the other hand the vertical components of the fountain flow act to reduce the suction pressures on lower surface of the configuration, and for side-by-side configurations with enough lateral spacing to produce a fountain, the type of

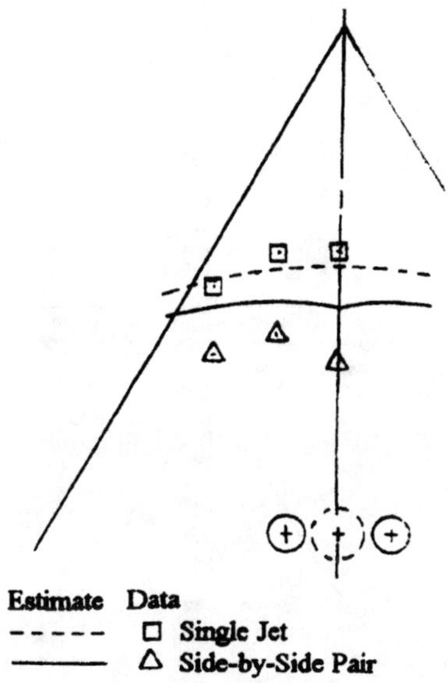

Fig. 4.5 Comparison of location of zero pressure line on the model lower surface for side-by-side pair with that for the equivalent single jet (from Ref. 12).

lower surface zero pressure line shown in Fig. 4.5 is produced. A review of the limited data available in Refs. 12 and 13 suggest that, for a side-by-side pair with sufficient lateral spacing, the distance from the jet to the zero pressure line is only about 80% that for an equivalent single jet.

Therefore, for the present method the distance X' from the jet pair to the most forward point of the zero pressure line, is assumed to be given by

$$\frac{X'}{d} = 0.48\,K_{\text{gb}}\left(\frac{S}{A_j}\right)^{0.2} V_e^{-0.4}\left(\frac{h}{d}\right)^{0.06V_e^{-0.7}} + \frac{h}{d}\tan(\delta - 90) \tag{4.3}$$

II. Ground Simulation

One of the critical issues in powered model testing is proper simulation of the ground environment. Reference 14 showed that excessively large lift losses were experienced in wind-tunnel test of jet flap configurations over a fixed ground board. A moving belt ground board (Fig. 4.7), with an endless belt moving at the same speed as the freestream, was developed to circumvent the problem. With a fixed ground board (top of Fig. 4.7) the forward-flowing wall jet from the impinging jet (or from the impinging jet sheet of a jet flap configuration) can progress further forward against the lower energy air in the boundary layer on the ground board than it would against the full velocity of the freestream.

The endless belt ground board serves two purposes. First, it eliminates the boundary layer that would exist on a fixed ground board. This is the purpose

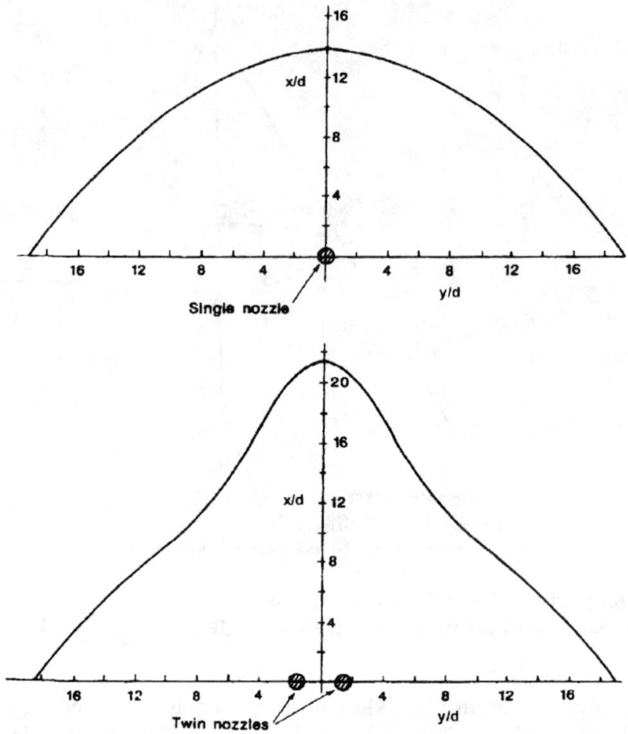

Fig. 4.6 Locations of zero pressure lines on the ground for single and twin nozzles based on flow visualization over a moving ground plane.[5]

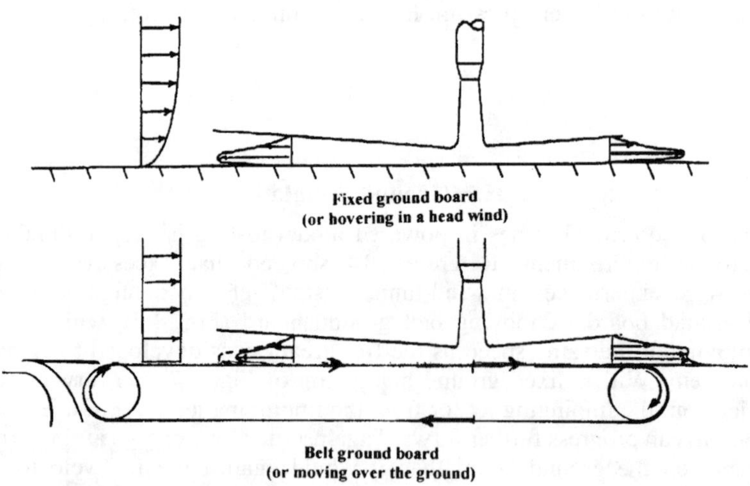

Fig. 4.7 Comparison of the wall jet profiles from a jet impinging on a fixed and on a moving ground plane.[14]

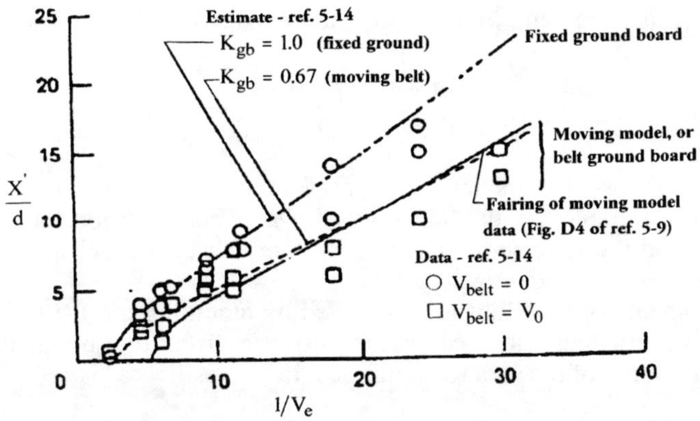

Fig. 4.8 Effect of type of ground simulation of the position of the zero pressure line at the configuration centerline.

cited in Ref. 14 and all of the early literature. To achieve this end, a slot is provided ahead of the belt (as shown at the left of the bottom sketch in Fig. 4.7) to remove the boundary layer at this point, and the belt is run at the same speed as the air flowing over it to prevent the redevelopment of the boundary layer.

The second, and equally important purpose of the belt, is to provide the scrubbing action the ground applies to the wall jet that is moving forward, with the aircraft, as it moves over the ground. This scrubbing action reduces the energy in the forward-flowing wall jet, as shown by the sketch at lower left in Fig. 4.7. Both of these two effects, eliminating the boundary layer on the ground and the erosion of the wall jets energy as a result of the scrubbing action of the ground surface moving aft under the forward-flowing wall jet, reduce its ability to penetrate forward against the freestream.

There have been several investigations and summarizations of the position of the ground vortex and of the forward penetration of the wall jet (Refs. 8–12 and 14–17). A summary of the most pertinent results of these several studies is presented in Fig. 4.8 (from Ref. 14). The ground vortex is very unsteady (Ref. 18), but, as Fig. 4.8 shows, the forward penetration of the zero pressure line for tests over a moving belt ground board is generally only about $\frac{2}{3}$ that over a fixed ground board.

III. Lift and Moment Estimates

The method for estimating the lift loss and pitching moments induced in STOL operation (transition in ground effect) for configurations with a single circular jet was developed in Ref. 10.

The lift loss induced at forward speed in ground effect can be expressed as

$$\frac{\Delta L}{T} = \left(\frac{\Delta L}{T}\right)_{GV} + \left(\frac{\Delta L}{T}\right)_{hov} + \left(\frac{\Delta L}{T}\right)_{wake} \tag{4.4}$$

and the pitching moment, like the lift, can be expressed as

$$\left(\frac{\Delta M}{Td_e}\right) = \left(\frac{\Delta M}{Td_e}\right)_{\mathrm{GV}} + \left(\frac{\Delta M}{Td_e}\right)_{\mathrm{hov}} + \left(\frac{\Delta M}{Td_e}\right)_{\mathrm{wake}} \tag{4.5}$$

where, in each case, the first term accounts for the effect of the ground vortex, the second term accounts for the hover suckdown effect modified for the effect of crossflow, and the third term accounts for the jet wake effect modified for the proximity of the ground.

The moment contributions are estimated by assuming the lift to be acting at an effective arm. In most cases the distance from the moment reference point to the center of the area for which the lift was estimated is used at the effective arm.

A. Ground Vortex Term

1. *Lift Estimates*

The ground vortex term is made up of four terms:

$$\left(\frac{\Delta L}{T}\right)_{\mathrm{GV}} = \left(\frac{\Delta L}{T}\right)_{\mathrm{GV,pos}} + \left(\frac{\Delta L}{T}\right)_{\mathrm{GV,neg}} + \left(\frac{\Delta L}{T}\right)_{\mathrm{us}} + \left(\frac{\Delta L}{T}\right)_{\mathrm{wb}} \tag{4.6}$$

where the first terms represents the effects of the positive pressures induced ahead of the zero pressure line (Fig. 4.9) and the second term represents the effects of the negative pressures induced between the zero pressure line and the jet. The shape and the position of the zero pressure line are calculated by Eqs. (4.1) and (4.2).

The lift increment as a result of the pressures in the positive pressure region is given by

$$\left(\frac{\Delta L}{T}\right)_{\mathrm{GV,pos}} = C_{\mathrm{p,GV,p}} \frac{A_{\mathrm{GV,p}}}{2A_j} \tag{4.7}$$

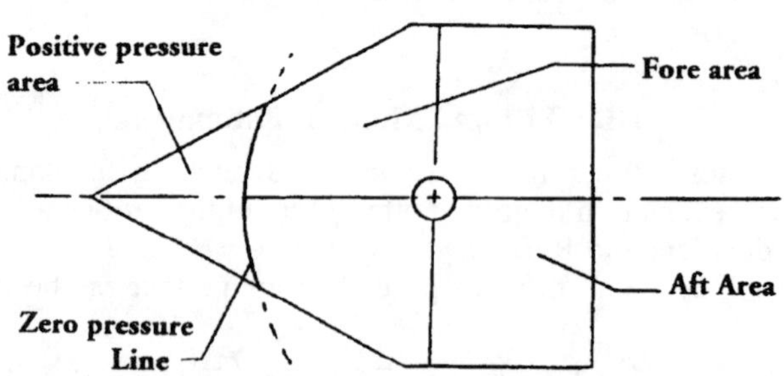

Fig. 4.9 Definition of areas used in the estimating method.

where $A_{\text{GV,p}}$ is the lower surface area forward of the zero pressure line and $C_{\text{p,GV,p}}$ is the average pressure coefficient in the positive pressure region and is given by

$$C_{\text{p,GV,p}} = K_{\text{gb}} \frac{(\delta/90)^2}{A_n^{0.25}} \frac{0.46 V_e f^{0.5}}{(h/d)(S_{\text{fwd}}/A_j)^{0.4}} \tag{4.8}$$

where A_n is the aspect ratio of the jet nozzle, A_j is the total jet area, f_p is the planform fineness ratio, and S_{fwd} is the planform area forward of the jet.

The lift increment caused by the negative pressures in the area between the zero pressure line and the jet is given by

$$\left(\frac{\Delta L}{T}\right)_{\text{GV,neg}} = C_{\text{p,GV,fore}} \frac{A_{\text{GV,fore}}}{2A_j} \tag{4.9}$$

where $A_{\text{GV,fore}}$ is the area between the zero pressure line and the jet and the average pressure on this area is the less negative of the pressures calculated by the two following expressions:

At the lower heights:

$$C_{\text{p,GV,fore}} = K_{\text{gb}} \frac{-10\, V_e}{A_n^{0.25}} \left(\frac{\delta}{90}\right)^2 \left(\frac{Y_{\text{ave}}}{d}\right)^{-2} \left(\frac{h}{d}\right)^{-8.4/\sqrt{S_{\text{fwd}}/A_j}} \tag{4.10}$$

At the higher heights:

$$C_{\text{p,GV,fore}} = K_{\text{gb}} \frac{-0.1 f_p^{0.25}}{A_n^{0.25}} \left(\frac{\delta}{90}\right)^2 \left(\frac{h}{d}\right)^{-2} \tag{4.11}$$

where Y_{ave} is the average width of the planform ahead of the jet station.

In addition to the pressures induced on the lower surface, the presence of the ground vortex forces the freestream to flow up and over itself putting the configuration in an upwash flowfield as shown in Fig. 4.10 (or downwash if the ground vortex centerline is ahead of the leading edge). The induced upwash induces lifting pressures on the upper surface. The method assumes that the upper surface increment is equivalent to the configuration operating at an increased angle of attack and that, for a body or a flat-plate configuration, the lift increment is given by

$$\left(\frac{\Delta L}{T}\right)_{\text{us}} = C_{L\alpha,\,\text{body}} \Delta\alpha_{\text{us}} \frac{S_{\text{ref}}}{2\Lambda_j} V_e^2 \tag{4.12}$$

where S_{ref} is the configuration reference area and the induced upwash angle is given by the following:

For negative values of X'' (leading edge ahead of the vortex centerline),

$$\Delta\alpha_{\text{us}} = K_{\text{gb}}^{0.2\left(\frac{X_{\text{jet}} - X_{\text{L.E.}}}{d}\right)} \left[0.7 - 0.7\frac{X''}{d} - 0.35\left[\frac{X''}{d}\right]^2\right] \frac{K_{\text{tgv}}}{V_e h/d} \frac{(\delta/90)^2}{A_n^{0.25}} \tag{4.13}$$

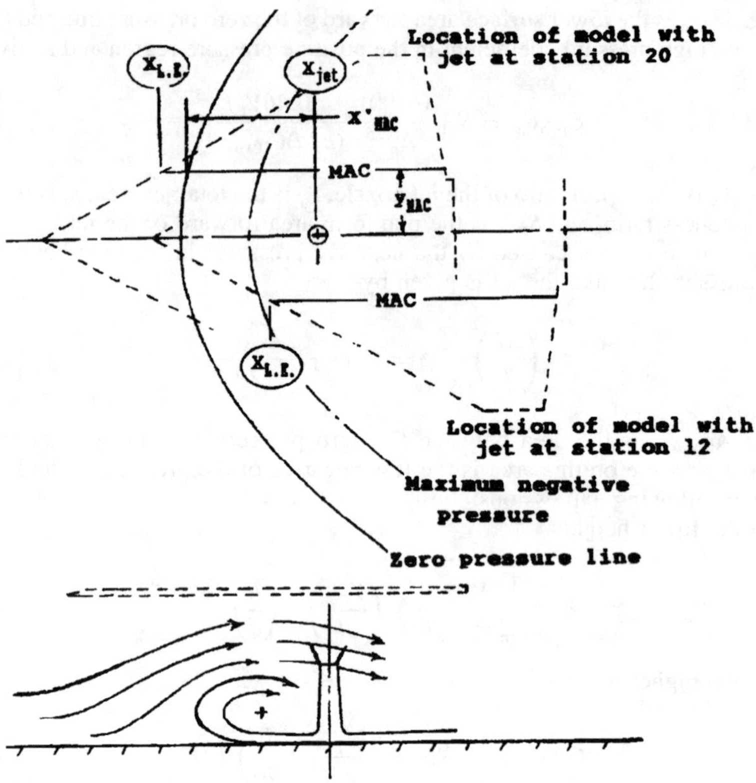

Fig. 4.10 Schematic of the effect of the ground vortex on the freestream flow (from Ref. 10).

For positive values of X'' (leading edge behind the vortex centerline),

$$\Delta\alpha_{us} = K_{gb}^{0.2\left(\frac{X_{jet}-X_{L.E.}}{d}\right)}\left[0.7 - 0.7\frac{X''}{d} - 0.35\left[\frac{X''}{d}\right]^2\right]\frac{K_{tgv}}{V_e h/d}\frac{(\delta/90)^2}{A_n^{0.25}} \tag{4.14}$$

where X'' is the distance from the assumed center of the ground vortex to the wing leading edge given by

$$\frac{X''}{d} = \frac{(X'_{mac}/2) - (X_{jet} - X_{L.E.})}{d} \tag{4.15}$$

and K_{tgv} accounts for the effects of the ground vortex being "trapped" under the configuration at low heights (Fig. 4.11). There is a critical height where the effect

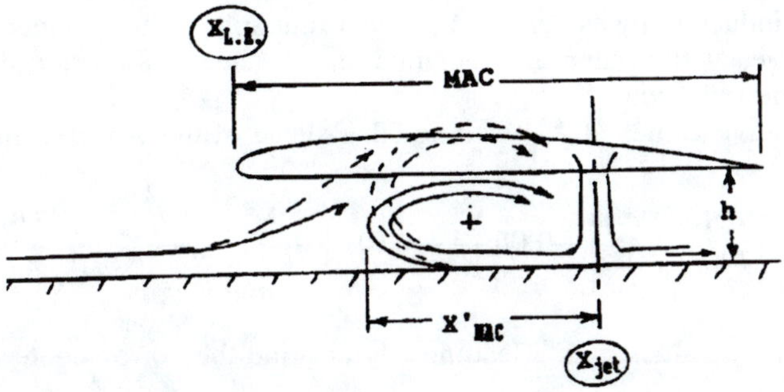

Fig. 4.11 Effect of trapping the ground vortex under the configuration at low heights.

of the trapped vortex changes. When the aircraft height is below the critical height

$$h = 0.5\sqrt{\frac{S}{A_j}}V_e X'_{\text{mac}}$$

$$K_{\text{tgv}} = \frac{h}{0.5\sqrt{S/A_j}V_e X'_{\text{mac}}} \tag{4.16}$$

and above

$$h = 0.5\sqrt{\frac{S}{A_j}}V_e X'_{\text{mac}}$$
$$K_{\text{tgv}} = 1.0 \tag{4.17}$$

The distance from the jet station to the vortex center is calculated at the spanwise station Y_{mac} of the mean aerodynamic chord and, accounting for the parabolic shape of the zero pressure line, is given by

$$X'_{\text{mac}} = X' - \frac{Y^2_{\text{mac}}}{4X'} \tag{4.18}$$

The preceding elements of the estimating method [Eqs. (4.4–4.18)] apply to a delta wing, or other configuration on which the entire lower surface is coplanar, and to the body of a wing-body configuration.

For a wing-body configuration the wing lift caused by the induced upwash is given by

$$\left(\frac{\Delta L}{T}\right)_{\text{wb}} = C_{L\alpha,\,\text{wing}}\Delta\alpha_{\text{wb}}\frac{S_{\text{ref}}}{2A_j}V_e^2 \tag{4.19}$$

where the induced upwash angle $\Delta\alpha_{\text{wb}}$ is a function of the distance of the wing leading edge, at the mean aerodynamic chord, ahead of the jet station and is given by the following:

For negative values of X'' (leading edge ahead of the vortex centerline),

$$\Delta\alpha_{\text{wb}} = K_{\text{gb}}^{0.2\left(\frac{X_{\text{jet}}-X_{\text{L.E.}}}{d}\right)}\left[-0.06\frac{X''}{d}+0.016\left(\frac{X''}{d}\right)^2\right]\frac{K_{\text{tgv}}}{V_e^2(h_w/d)^a}\frac{(\delta/90)^2}{A_n^{0.25}} \tag{4.20}$$

For positive values of X'' (leading edge behind the vortex centerline),

$$\Delta\alpha_{\text{wb}} = K_{\text{gb}}^{0.2\left(\frac{X_{\text{jet}}-X_{\text{L.E.}}}{d}\right)}\left[-0.06\frac{X''}{d}\right]\frac{K_{\text{tgv}}}{V_e^2(h_w/d)^a}\frac{(\delta/90)^2}{A_n^{0.25}} \tag{4.21}$$

where the exponent a is given by

$$a = 1 + 0.06V_e\left(\frac{X_{\text{jet}} - X_{\text{L.E.}}}{d}\right)^2 \tag{4.22}$$

If the wing lower surface is not coplanar with the body lower surface, then the height of the wing above the ground h_w is used for h in calculating the upwash [Eqs. (4.20) and (4.21)].

At some combinations of low velocity ratio V_e and low heights h_w, Eq. (4.20) will calculate increments of induced upwash angle of attack that would carry the wing beyond stall. For these conditions it is suggested that if

$$\Delta\alpha_{\text{wb}} > (\alpha_{\text{stall}} - \alpha)$$

then use

$$\Delta\alpha_{\text{wb}} = (\alpha_{\text{stall}} - \alpha) \tag{4.23}$$

2. *Pitching-Moment Estimates*

The ground vortex term is made up of four terms:

$$\left(\frac{\Delta M}{Td}\right)_{\text{GV}} = \left(\frac{\Delta M}{Td}\right)_{\text{GV,pos}} + \left(\frac{\Delta M}{Td}\right)_{\text{GV,neg}} + \left(\frac{\Delta M}{Td}\right)_{\text{us}} + \left(\frac{\Delta M}{Td}\right)_{\text{wb}} \tag{4.24}$$

The first term, the moment caused by the positive lift generated forward of the zero pressure line, is given by

$$\left(\frac{\Delta M}{Td}\right)_{\text{GV,p}} = \left(\frac{\Delta L}{T}\right) = \frac{(X_{\text{C.G.}} - X_p)}{d} \tag{4.25}$$

For the second term, negative lift generated between the jet and the zero pressure line, the moment contribution is given by

$$\left(\frac{\Delta M}{Td}\right)_{\mathrm{GV,p}} = \left(\frac{\Delta L}{T}\right)\frac{(X_{\mathrm{C.G.}} - X_{\mathrm{GV,fore}})}{d} \tag{4.26}$$

The upwash-induced lift increment is generated in a curved flowfield (Fig. 4.9), which produces a camber-type loading. It is therefore assumed that the lift is applied at the midchord point of the mean aerodynamic chord of the planform or wing. For the method the moment caused by the upper surface lift increment is given by

$$\left(\frac{\Delta M}{Td}\right)_{\mathrm{us}} = \left(\frac{\Delta L}{T}\right)_{\mathrm{us}}\frac{(X_{\mathrm{C.G.}} - X_{0.5\,\mathrm{MAC,body}})}{d} \tag{4.27}$$

and the moment caused by the upwash-induced lift on the exposed wing is given by:

$$\left(\frac{\Delta M}{Td}\right)_{\mathrm{wb}} = \left(\frac{\Delta L}{T}\right)_{\mathrm{wb}}\frac{(X_{\mathrm{C.G.}} - X_{0.5\,\mathrm{MAC,wing}})}{d} \tag{4.28}$$

B. Hover Suckdown Term

1. *Lift Estimates*

The hover suckdown term is made up of three terms:

$$\left(\frac{\Delta L}{T}\right)_{\mathrm{hov}} = \left(\frac{\Delta L}{T}\right)_{\infty} + \left(\frac{\Delta L}{T}\right)_{\mathrm{hov,body}} + \left(\frac{\Delta L}{T}\right)_{\mathrm{hov,wing}} \tag{4.29}$$

where $(\Delta L/T)_{\infty}$ is the lift loss induced out-of-ground effect as estimated in Sec. I of Chapter 2 [Eq. (2.2) or (2.4)].

The current method assumes that the pressures induced in hover by the impingement of the jet (Sec. II of Chapter 2), are not altered by the crossflow, but their effects are constrained to the region of the lower surface aft of the zero pressure line as shown in Fig. 4.8. The hover suckdown term therefore is the sum of the lift loss induced forward of the jet (between the jet and the zero pressure line) and that induced aft of the jet, and for a body or flat-plate configuration the lift increment is given by

$$\left(\frac{\Delta L}{T}\right)_{\mathrm{hov,body}} = C_{\mathrm{p,hov}}\left(\frac{A_{\mathrm{GV,fore}} + A_{\mathrm{GV,aft}}}{2A_j}\right)_{\mathrm{body}} \tag{4.30}$$

and the lift induced on the exposed wing is given by

$$\left(\frac{\Delta L}{T}\right)_{\text{hov,wing}} = C_{\text{p,hov}}\left(\frac{A_{\text{GV,fore}} + A_{\text{GV,aft}}}{2A_j}\right)_{\text{wing}} \tag{4.31}$$

where the pressure induced in hover is given by

$$C_{\text{p,hov}} = \frac{\sin^2\delta}{1 + V_e(A_{n-1})}\left(\frac{h}{D_p - d}\right)^e K_{\text{sj}}K_{\text{tv}} \tag{4.32}$$

where h is the height of the lower surface of the body for the body term [Eq. (4.30)], and for the wing term [Eq. (4.31)] h is the height of the lower surface of the exposed wing. The constant K_{sj} is estimated by Eq. (2.11), the exponent e is estimated by Eq. (2.12), the critical height $(h/d)_{\text{tv}}$ is estimated by Eq. (2.8), and in the trapped vortex region K_{tv} is estimated by the following:

Below $(h/d)_{\text{tv}}$:

$$K_{\text{tv}} = 1 - \left(1 - \frac{h/d}{(h/d)_{\text{tv}}}\right)^{1.66} \tag{4.33}$$

Above $(h/d)_{\text{tv}}$:

$$K_{\text{tv}} = 1.0 \tag{4.34}$$

2. *Pitching-Moment Estimates*

The hover moment term is made up of three terms:

$$\left(\frac{\Delta M}{Td}\right)_{\text{hov}} = \left(\frac{\Delta M}{Td}\right)_{\infty} + \left(\frac{\Delta M}{Td}\right)_{\text{hov,body}} + \left(\frac{\Delta M}{Td}\right)_{\text{hov,wing}} \tag{4.35}$$

where $(\Delta M/Td)_\infty$ is the moment induced out-of-ground effect as estimated in Sec. I of Chapter 2 [Eq. (2.7)].

The pitching moment induced in hover, for single jet configurations, is, as noted in Sec. II.A of Chapter 2, the difference between the nose-down moment generated by the suckdown ahead of the jet and the nose-up moment induced aft of the jet. At higher heights, as shown in Ref. 19, these moments are given by the lift loss acting at the center of area. However below a height equal to the distance from the jet to the center of area, the moment arm X_e reduces rapidly.

Forward of the jet (between the zero pressure line and the jet) and above a height given by $h = X_{\text{C.G.}} - X_{\text{fore}}$,

$$X_{\text{e,fore}} = (X_{\text{C.G.}} - X_{\text{fore}}) \tag{4.36}$$

and below $h = X_{\text{C.G.}} - X_{\text{fore}}$,

$$X_{\text{e,fore}} = (X_{\text{C.G.}} - X_{\text{fore}})\left\{1 - \left[1 - \frac{h}{(X_{\text{C.G.}} - X_{\text{fore}})}\right]^2\right\} \tag{4.37}$$

aft of the jet and above a height given by $h = |X_{C.G.} - X_{aft}|$,

$$X_{e,aft} = (X_{C.G.} - X_{aft}) \tag{4.38}$$

and below a height given by $h = |X_{C.G.} - X_{aft}|$,

$$X_{e,aft} = (X_{C.G.} - X_{aft})\left\{1 - \left[1 - \frac{h}{-(X_{C.G.} - X_{aft})}\right]^2\right\} \tag{4.39}$$

And the moment caused by hover suckdown effects on the body is given by

$$\left(\frac{\Delta M}{Td_e}\right)_{hov,body} = C_{p,hov}\left[\left(\frac{A_{GV,fore}}{2A_j}\right)\left(\frac{X_{e,fore}}{d}\right) + \left(\frac{A_{GV,aft}}{2A_j}\right)\left(\frac{X_{e,aft}}{d}\right)\right] \tag{4.40}$$

where the fore and aft areas and their corresponding effective arms are those for the body elements.

Similarly the moment caused by hover suckdown effects on the wing is given by

$$\left(\frac{\Delta M}{Td_e}\right)_{hov,wing} = C_{p,hov}\left[\left(\frac{A_{GV,fore}}{2A_j}\right)\left(\frac{X_{e,fore}}{d}\right) + \left(\frac{A_{GV,aft}}{2A_j}\right)\left(\frac{X_{e,aft}}{d}\right)\right] \tag{4.41}$$

where the fore and aft areas and their corresponding effective arms are those for the wing elements.

C. Jet Wake Term

1. *Lift Estimates*

When close to the ground, the jet wake is truncated (Fig. 4.2), and the suction pressures induced by the jet/freestream interaction in the region aft of the jet are reduced.

The jet wake term is given by

$$\left(\frac{\Delta L}{T}\right)_{wake} = \left(\frac{\Delta L}{T}\right)_j + \left(\frac{\Delta L}{T}\right)_\Gamma + \left(\frac{\Delta L}{T}\right)_{GV,aft} \tag{4.42}$$

where the first two terms are the lift loss and lift gain (if any) induced in transition, out-of-ground effect and estimated by Eqs. (3.24) and (3.26), respectively. Only the body term is adjusted for the effects of ground proximity by

$$\left(\frac{\Delta L}{T}\right)_{GV,aft} = \Delta C_{p,GV}\frac{S_{aft}}{2A_j} \tag{4.43}$$

The increment of pressure coefficient $\Delta C_{p,GV}$ is the difference between the pressure coefficient induced in transition, in-ground effect, and that induced out-of-ground effect. This pressure coefficient increment was determined in

Ref. 4 and is given by

$$\Delta C_{\text{p,GV}} = K_{\text{GV}} 0.05\, V_e A_n^{0.5} \left(\frac{\delta}{90}\right)^2 \left(\frac{h}{d}\right)^{-1.5} \tag{4.44}$$

2. *Pitching-Moment Estimates*

The pitching moment caused by the jet wake effects, for single jet configurations, is given by

$$\left(\frac{\Delta M}{Td}\right)_{\text{wake}} = \left(\frac{\Delta M}{Td}\right)_{\text{t,body}} + \left(\frac{\Delta M}{Td}\right)_{\text{t,wing}} + \left(\frac{\Delta L}{Td}\right)_{\text{GV,aft}} \tag{4.45}$$

where $(\Delta M/Td)_{\text{t,body}}$ and $(\Delta M/Td)_{\text{t,wing}}$ are estimated by Eqs. (3.44)–(3.46), and the moment increment caused by the ground vortex $(\Delta M/Td)_{\text{GV,aft}}$ is given by

$$\left(\frac{\Delta M}{Td}\right)_{\text{GV,aft}} = \left(\frac{\Delta L}{T}\right)_{\text{GV,aft}} \frac{(X_{\text{C.G.}} - X_{\text{GV,aft}})}{d} \tag{4.46}$$

IV. Fountain Effects

The lift and moment estimation method presented in the preceding section is applicable only to single jet configurations. A comparable method applicable to multijet configurations is not available. However the effects of crossflow on the fountain are presented in Ref. 12. Shown in Fig. 4.12 are three tandem jet configurations [1) the jet at station 12 and the jet immediately downstream; 2) the jet at station 12 and the jet at station 20; and 3) the jet at station 12 and the jet furthest downstream] and one side-by-side jet configuration (the two side-by-side jets at station 12), which were evaluated.

A typical centerline pressure distribution (Fig. 4.13) for an intermediate jet-spacing-to-diameter ratio shows the effects of the ground vortex generated forward of the front jet in producing the positive pressures forward of station −12 and the negative pressures in the ground vortex region (stations −4 to −12). With crossflow ($V_e = 0.15$) the positive pressure induced by the fountain is reduced to nearly zero, and the suckdown pressure between the front jet and the fountain is also significantly reduced. In addition the effects of crossflow in increasing the negative pressures in the wake of the jets ($x > 5$) is also apparent.

The effects of crossflow on the shape and size of the positive pressure region induced by the impingement of the fountain are shown in Fig. 4.14. Crossflow reduces the size of the positive pressure region but does not change its location. It remains essentially centered between the jets. Note also that the aft curvature of the lines connecting the peak positive pressures in the impingement region and connecting the peak suckdown pressures between the jets and the fountain is caused by the width of the planform increasing with distance aft. This curvature is not significantly affected by the crossflow.

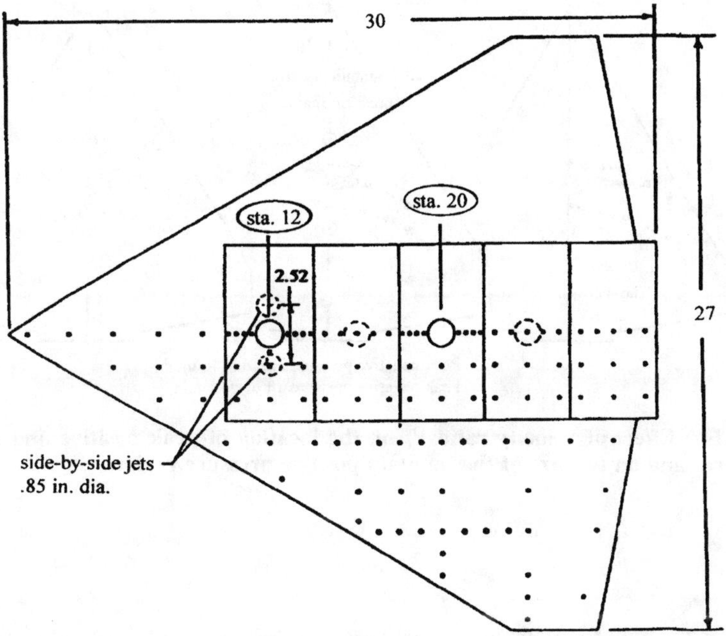

Fig. 4.12 Configuration of the flap-plate model showing the locations of the jets and pressure taps.

The lift induced by the fountain was obtained by integrating the pressures over the positive pressure region for each height and velocity ratio tested. Typical results for two heights are shown in Fig. 4.15 as ratios of the integrated lift increment to the corresponding increment obtained in hover. The increments are small, and there is therefore considerable scatter in the data, but these and corresponding plots at other heights clearly show that the fountain increment decreases essentially linearly with V_e.

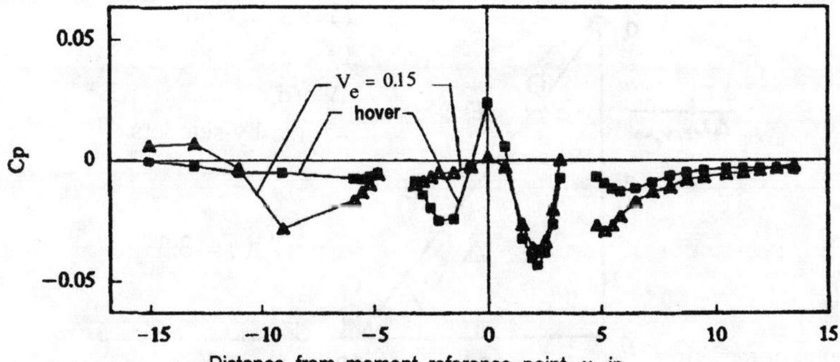

Fig. 4.13 Typical centerline pressure distribution, jets at stations 12 and 20, $h/d_e = 1.7$.

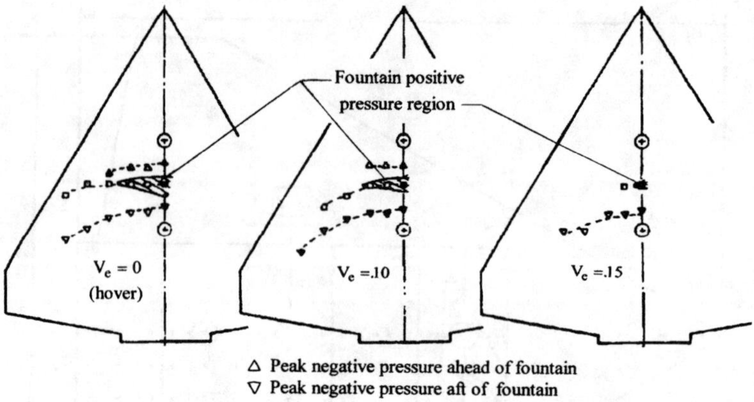

Fig. 4.14 Effect of velocity ratio V_e on the location of peak positive and negative pressures and on the size of the fountain positive pressure region.

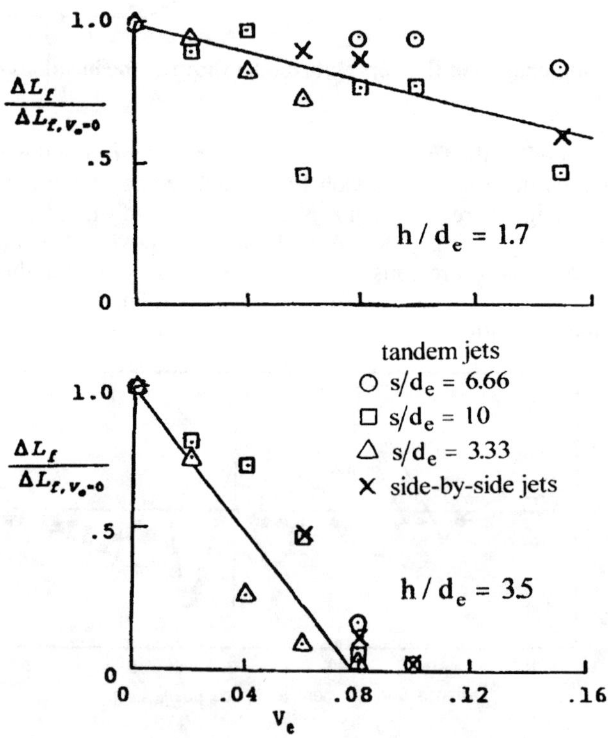

Fig. 4.15 Effect of height and V_e on fountain lift.

Nomenclature

A = aspect ratio of planform or element of configuration under consideration
$A_{GV,aft}$ = planform area aft of the jet, ft^2
$A_{GV,fore}$ = planform area between zero pressure line and jet, ft^2
$A_{GV,p}$ = planform area forward of the zero pressure line, ft^2
A_j = jet exit area; total area unless otherwise noted, ft^2
A_n = aspect ratio of jet nozzle
a = exponent used in estimating wing-body upwash [Eq. (4.22)]
C_{L_α} = power-off lift-curve slope
C_p = pressure coefficient, $= \Delta P/q_j$
D_p = equivalent diameter of planform area, ft
d = diameter of individual jet, ft
d_e = equivalent diameter of total jet area, ft
exp = exponent used in estimating hover suckdown pressure [Eq. (2.12)]
f_p = planform fineness ratio
h = height of body surface above ground, ft
h_w = height of wing above ground, ft
K_{gb} = factor accounting for ground condition or type of ground simulation (see Fig. 4.8)
K_{sj} = factor used in estimating hover suckdown increment [Eq. (2.11)]
K_{tgv} = adjustment factor for effect of trapping the ground vortex at low heights [Eq. (4.16)]
K_{tv} = adjustment factor for effect of trapped vortex in hover [Eqs. (4.31) and (4.33)]
mac = mean aerodynamic chord, ft
q = freestream dynamic pressure, lb/ft^2
q_j = jet dynamic pressure at nozzle, lb/ft^2
S = total planform area of configuration, or area of part of configuration under consideration, ft^2
S_{ref} = reference area used in calculation of coefficients, ft^2
T = total jet thrust, lb
V_e = effective velocity ratio, $\equiv \sqrt{q/q_e}$
X = distance from the geometric center of an area to the moment reference point, ft
$X_{C.G.}$ = station at which the moment reference point is located, ft
X_e = effective arm at which jet-induced lift increment acts, ft
X_{jet} = station at which the jet is located (see Fig. 4.10), ft
$X_{L.E.}$ = station at which the leading edge of mean aerodynamic chord is located (see Fig. 4.10), ft
X_{pos} = station at which the geometric center of area of the positive pressure region forward of the zero pressure line is located, ft
X' = longitudinal distance of zero pressure point, on model centerline, ahead of jet (see Fig. 4.2) [Eqs. (4.1) and (4.3)], ft
X'_{mac} = longitudinal distance of zero pressure line at lateral station of mean aerodynamic chord (see Fig. 4.10), ft
x = longitudinal distance from jet station, ft

Y_{ave} = average width of the planform ahead of jet station, ft
y = lateral station from the configuration centerline, ft
y_{mac} = lateral distance of the mean aerodynamic chord from the configuration centerline (see Fig. 4.10), ft
α = angle of attack, deg
ΔM = pitching-moment increment, ft · lb
ΔP = jetinduced increment of pressure, lb/ft^2
ΔL = lift-loss increment, lb
$\Delta\alpha$ = upwash angle induced by the ground vortex [Eqs. (4.13) and (4.14)]
δ = jet deflection angle, measured from the horizontal, deg

Subscripts

aft = region aft of the jet
body = body of a wing-body configuration or a flatplate configuration
C.G. = center of gravity, or moment reference point
data = experimental data
f, or fwd = front or forward of the jet station, or fountain increment
fore = region between the zero pressure line and the jet station
G = ground
GV = ground vortex contribution
hov = hover suckdown contribution
j = jet
neg = negative pressure region
p, or pos = positive pressure region
sj = single jet
tv = trapped vortex condition
us = upper surface contribution
wake = jet wake contribution
wb = wing body
wing = wing

References

[1]Abbott, W. A., "Studies of Flow Fields Created by Vertical and Inclined Jets Moving over a Horizontal Surface," ACR Cp no. 911, 1964.

[2]Colin, P. E., and Olivari, D., "The Impingement of Circular Jet Normal to a Flat Surface with and Without Cross Flow," von Karman Inst. of Fluid Dynamics, Rept. AD688953, Rhode-St. Genese, Belgium, Jan. 1969.

[3]Schwantes, E., "The Recirculation Flow Pattern of a VTOL Lifting Engine," NASA TT F-14912, June 1973.

[4]Weber, H. A., and Gay, A., "VTOL Reingestion Model Testing of Fountain Control and Wind Effects," Prediction Methods for V/STOL Propulsion Aerodynamics, Naval Postgraduate School, Monterey, CA, July 1975, pp. 358–380.

[5]Knowles, K., Bray, D., Bailey, P. J., and Curtis, P., "Impinging Jets in Cross Flow," *The Aeronautical Journal of the Royal Aeronautical Society*, Vol. 92, Feb. 1992, pp. 47–56.

[6]Knowles, K., and Bray, D., "Ground Vortex Formed by Impinging Jets in Crossflow," *Journal of Aircraft*, Vol. 30, No. 6, 1993, pp. 872–878.

[7]Knowles, K., and Bray, D., "Computation of Normal Impinging Jets in Cross-Flow and Comparison with Experiments," *International Journal for Numerical Methods*, Vol. 13, No. 10, 1991, pp. 1225–1233.

[8]Stewart, V. R., and Kuhn, R. E., "A Method for Estimating the Propulsion Induced Aerodynamic Characteristics of STOL Aircraft in Ground Effect," NADC 80226-60, Naval Air Development Center, Warminster, PA, Aug. 1983.

[9]Stewart, V. R., and Kuhn, R. E., "Estimation of Lift and Pitching Moment Induced on Jet STOVL Aircraft by the Ground Vortex," WL-TR-93-3061, Aug. 1993.

[10]Kuhn, R. E., "An Analysis of the Pressures, Forces and Moments Induced by the Ground Vortex Generated by a Single Impinging Jet," NASA CR-4765, Feb. 1997.

[11]Eyles, J., Lawson, N., and Knowles, K., "A Study Using PIV and LDA of a Compressible STOVL Ground Vortex Flow," AIAA International powered Lift Conference, AIAA Paper 2002-5960, Williamsburg, VA, Nov. 2002.

[12]Kuhn, R. E., "The Effect of Crossflow on the Pressures and Lift Induced by the Fountain Generated Between Two Impinging Jets," *Proceeding of the Royal Aeronautical Society International Powered Lift Conference*, London, England, Sept. 1998, pp. 27.1–27.11.

[13]Barata, J., M. M., "Fountain Flows Produced by Multiple Impinging Jets in a Crossflow," 33rd Aerospace Sciences Meeting and Exhibit, AIAA Paper 95-0190, Reno, NV, Jan. 1995.

[14]Turner, T. R., "A Moving-Belt Ground Plane for Wind-Tunnel Ground Simulation and Results for Two Jet-Flap Configurations," NASA TN-D4228, Nov. 1967.

[15]Kuhn, R. E., Del Frate, J. H., and Eshleman, J. E., "Ground Vortex Flow Field Investigation," *1987 Ground Vortex Workshop*, NASA CP-10008, NASA Ames Research Center, April 1987, pp. 61–90.

[16]Paulson, J. W., and Kemmerly, G. T., "An Assessment of Ground Vortex Effects Determined by Static and Dynamic Testing Techniques," *1987 Ground Vortex Workshop*, NASA CP-10008, Proceedings of Meeting at NASA Ames Research Center, Moffett Field, CA, April 1987, pp. 121–146.

[17]Steward, V. R., Kuhn, R. E., and Walters, M. M., "Characteristics of the Ground Vortex Developed by Various V/STOL Jets at Forward Speed," AIAA Aircraft Design, System and Technology Meeting, AIAA Paper 83-2494, Ft. Worth, TX, Oct. 1983.

[18]Billet, M. L., and Cimbala, J. M., "Summary of an Experimental Investigation of the Ground Vortex," *1987 Ground Vortex Workshop*, NASA CP-10008, Proceedings of Meeting at NASA Ames Research Center, Moffett Field, CA, April 1987, pp. 39–60.

[19]Kuhn, R. E., Stewart, V. R., and Wardwell, D. A., "Estimation of Lift and Pitching Moment Induced on Jet STOVL Aircraft Hovering in Ground Effect," WL-TR-93-3046, Wright Laboratory, A. F. Material Command, Wright Patterson Air Force Base, Ohio, Aug. 1993.

Chapter 5

Hot-Gas Ingestion

I. Overview

THE INGESTION by a gas turbine engine, of air that is hotter than the surrounding ambient air has a number of consequences. The most immediate is that the overall pressure rise, which is proportional to the nondimensional rotor speed $N/\sqrt{T}$, is reduced, which reduces the thrust of the engine. This will usually be countered by either the pilot or the engine control system increasing the fuel to maintain thrust, with the consequence that the actual rotor speed and the turbine inlet temperature [or jet pipe temperature (JPT)] increase. The greater the temperature rise as a result of ingestion, the greater the increases needed to maintain thrust. This can only be accommodated up to the point where either the maximum rotor speed limit or the maximum turbine inlet temperature limit is reached. If the temperature limit is reached, then fuel flow actually has to be reduced to maintain the temperature, reducing the actual rotor speed, and thus reducing the nondimensional speed even further. Figure 5.1 illustrates how thrust is lost and temperature increased as a result of ingestion.

Hot-gas ingestion (HGI) can develop very quickly, with the temperature of the air entering the inlets fluctuating as a function of height and of time. The distribution of temperature reaching the engine face can also vary quite markedly, with the variation referred to as a temperature distortion. Figure 5.2 shows two plots of data taken during a model test vertical landing. The top plot shows both the variation of temperature at a single point on the engine face (probe 3) and the mean engine face temperature, from a total of 45 measurement points (thermocouples). In this case the individual probe has lower ingestion than the overall mean. The bottom plot shows how the data from the 45 thermocouples translate into a measure of the distortion. Fluctuations of this sort make it difficult to document the average temperature rise, whereas extreme values of distortion can lead to compressor stall and possible engine failure.

HGI, then, can be a limiting factor in determining the maximum vertical landing mass of the aircraft at the highest operating temperatures required by the operator. HGI has been the subject of many investigations, as the 46 references in this chapter show. The general flow phenomena are well understood,[1–8] but our ability to predict inlet temperature rise is poor. This is mostly because of the unsteady and unstable nature of the flow phenomena involved.

The flow mechanisms leading to hot-gas ingestion (Fig. 5.3) can be broken into three categories: near field, crossflow or midfield, and farfield. Near-field

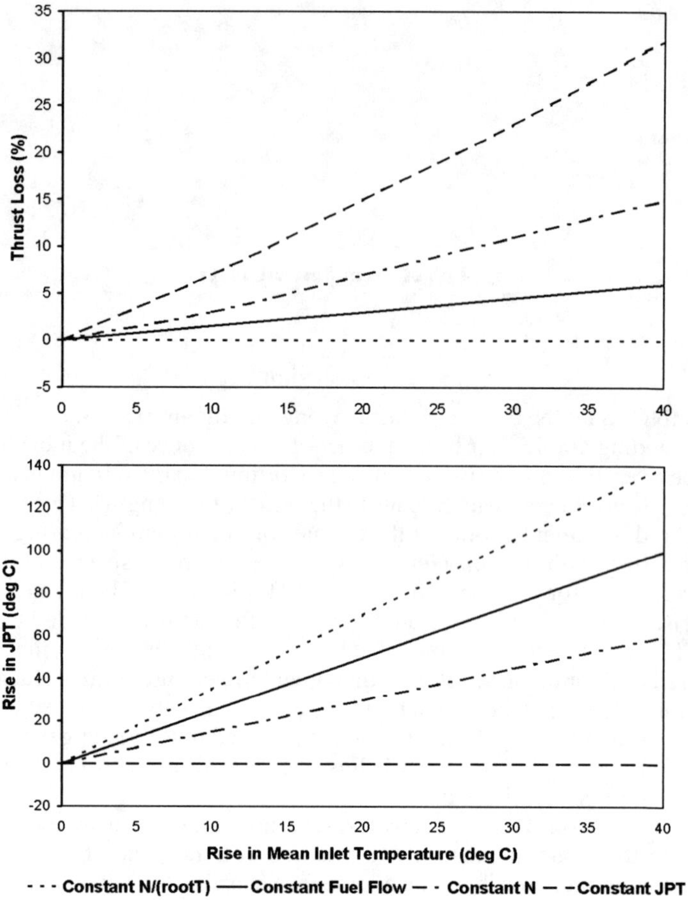

Fig. 5.1 Effect of mean inlet temperature rise on thrust.[1]

ingestion is usually the most serious and occurs when fountain flows find their way to the aircraft inlet. This can either be by their impingement on the underside of the fuselage, from where they flow towards the inlets, or by direct ingestion of the fountain into the inlet. Because of the short path length of the flow from the nozzle, there is relatively little mixing of the jet, and the temperatures are consequently quite high. Distortion can also be high if the ingestion is directed to only one portion of the inlet.

Far-field ingestion comes when the outward flowing wall jets lose enough driving force, in the form of dynamic pressure, that the buoyancy force as a result of the jet temperature (the reduced density of the hot gas) causes the wall jets to separate from the ground (Fig. 5.4). From here they are influenced by the entrainment flowfield and by any wind there might be. The long path length of this ingestion means the flow is well mixed by the time it reaches the

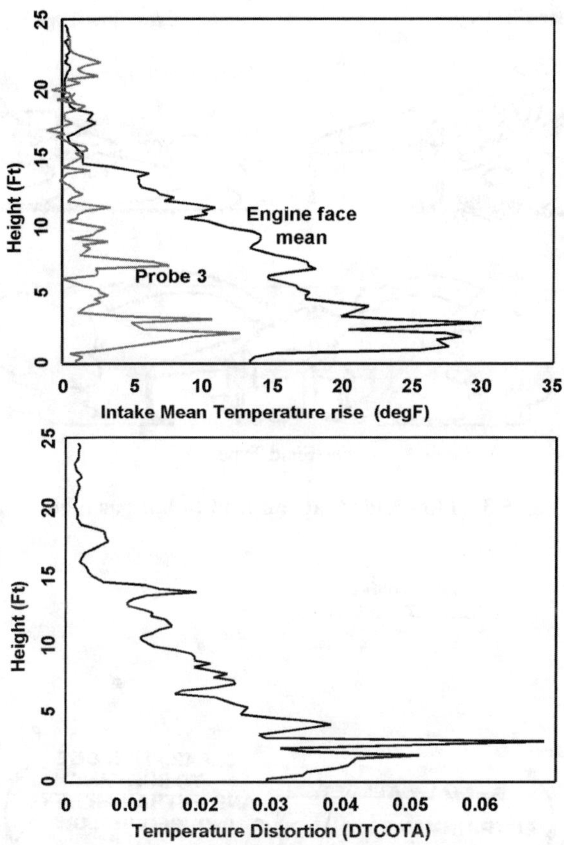

Fig. 5.2 Typical thermocouple signals during landing showing temperature fluctuation.[1]

inlet, so that temperatures and distortion are generally quite low. On still days the Harrier will experience this ingestion during a prolonged hover at 50–80 ft—there being no wind to disperse the warm gas rising from the ground.

Midfield, or crossflow, ingestion comes as a consequence of the instability of jet and fountain flows. The action of entrainment into the flow leads to discrete vortices, or eddies, which can detach from the main jet. In fact the edge of a ground jet consists of a whole series of eddies, of varying sizes, traveling in a streamwise direction. The neat jet boundaries shown in diagrams are notional averages and far removed from the reality of the time-dependent flow. Once detached from either a ground jet or fountain, these eddies, containing hot gas, are subjected to a number of different forces. They have their initial momentum, buoyancy forces, the entrainment flowfield, and any wind. The process defining how big the eddies grow, where they detach, and where they go after this is chaotic, but is considered to be stochastic, that is, it can be described statistically.

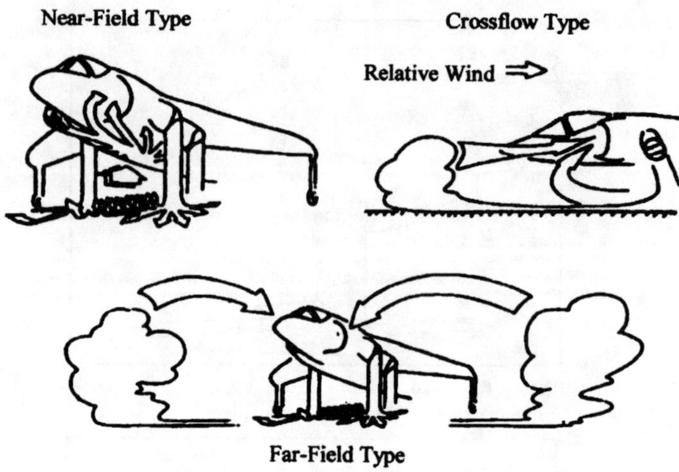

Fig. 5.3 Flowfields that can lead to hot-gas ingestion.

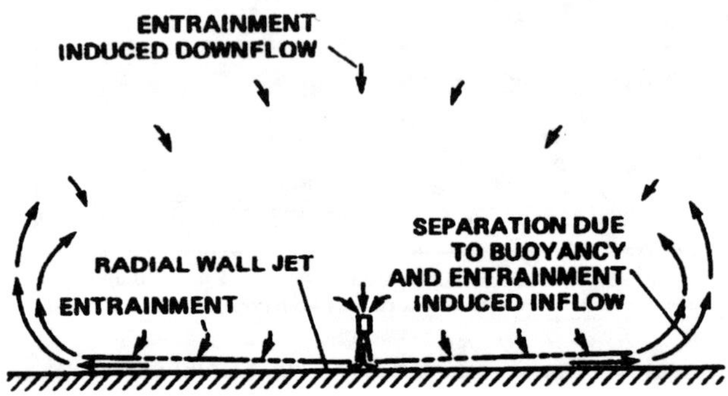

Fig. 5.4 Downflow induced by the entrainment action of the radial wall jets.

Midfield ingestion tends to be the most troublesome for the designer because of its unpredictability.

II. Fountain Characteristics

A. Closely Spaced Jets

The action of entrainment into jets causes them to both spread and slow down. Spreading rates (as an effective jet edge angle) can be found in the literature, although it is difficult to find values for anything other than circular jets. The reason for mentioning this is that for closely spaced jets there will be a downstream distance at which the jets will merge and will fail to produce a fountain when they impinge on the ground. To be fully merged, a cross-sectional velocity profile would be indistinguishable from a single jet, but for the purposes of fountain suppression it is enough that two separate ground jets are prevented

Fig. 5.5 Front nozzles splayed inward.[9]

from forming. In fact, there will be a case where a very weak fountain is formed, but it is totally entrained back into the jets.

With closely spaced jets, such as those on the Harrier, the height above ground, above which the jets do not merge and a fountain is formed, is really quite low. Given that fountain flows are the main cause of HGI problems, this might be thought to be a good thing, but the downside is that once formed fountains from closely spaced jets are very strong and energetic. Generally when the jets are close together, the fountain is stronger and forms at a lower height.

The strength of the fountain between two jets can be reduced by vectoring, or splaying, the jets away from each other. This has the byproduct of increasing the height at which the fountain forms. Conversely, by vectoring or splaying towards each other, the fountain formation is delayed to a lower height, but the strength is increased.

The Shoeburyness* Harrier (Fig. 5.5; Ref. 9) is a good example of the latter approach. For this full-scale rig test of a Pegasus engine with afterburning units in the front nozzles, known as plenum chamber burning (PCB), the front nozzles were splayed in by a full 15 deg each. It was predicted from earlier sub-scale tests that complete elimination of the front fountain, and its very high temperatures, was the only way to prevent engine stall from HGI.

The angle that the fountain makes to the vertical needs to be considered. Between two similar ground jets it will of course be vertical, as a time average. If, though, one of the ground jets has significantly more depth, or a different velocity vs height profile, then the resulting fountain will have a vector component in the same direction as the deeper jet.

Fountain flows from closely spaced jets are much more likely to impinge on the aircraft, in the region of the nozzles, before finding their way to an inlet,

---

*Shoeburyness in the United Kingdom was the location of the test site where the airframe/engine was tested.

rather than causing direct ingestion. This is a direct consequence of the low height of fountain formation. This makes these fountain flows amenable to control by airframe-mounted devices, such as strakes and dams.

The Harrier[9] displays most of the characteristics described in this section and some additional ones too. The jets are closely spaced and grouped around the center of gravity. The lateral pairs of jets are similar, but not the fore and aft pairs. The front nozzles have a 5-deg outward splay angle, and the rear nozzles, which are more closely spaced than the front, have a 12-deg outward splay angle when vectored to the hover position. These values were arrived at for reasons other than HGI, but they are fortuitously close to the optimum.

The close spacing means that no fountains are formed until the last 10 ft or so of a descent. The rear nozzles would be expected to produce a stronger fountain than the front because of their spacing, but the increased splay weakens this enough that the fountain core formed by all four jets, and containing hot rear jet gases, is not angled forwards. If anything, it is angled slightly aft. Figure 5.6 shows a sketch of the lift-improvement-device (LID) installation on the AV-8B Harrier. The gun-pod installation on the Harrier was responsible for collecting the fountain flow striking the fuselage and channelling it forward into the engine inlets. Most of this came from the front nozzles, but, although relatively cool, the flow could still give problems with HGI. The fence at the front of the gun pods was intended to block the path of this gas, which it

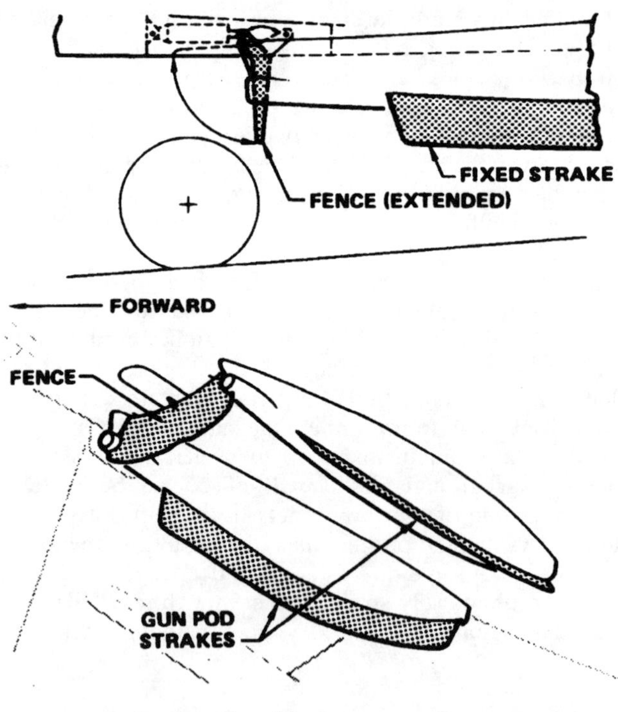

Fig. 5.6 LID installation on the AV8B Harriers.

did adequately, and incidentally gave a big boost to the fountain lift felt by the aircraft close to the ground. This installation, however, was not the completely optimized low HGI solution. Some HGI was caused either by gas spilling over the fence or simply impinging the fuselage forward of the fence. Reference 8 describes the design and verification of a fix for this, which was to add more longitudinal strakes forward of the fence. This fix never went onto the production aircraft.

B. Widely Spaced Jets

For widely spaced jets there is little likelihood of lift jets merging at typical hover heights, so that there are always likely to be multiple ground jets of varying strengths.[10] However, at a high enough height the fountain flows will be weak and will be entrained back into the jets. The fountain strengthens as height decreases, and there will be a height at which the fountain is strong enough to reach the aircraft. This can be as high as 50 ft for some configurations. The angle the fountain makes to the vertical is very important in this instance as it is possible for the fountain to be ingested directly into the inlet without impinging on the fuselage.

The wide spacing of the jets means that at low height the fountain will be weaker than it would be for closely spaced jets. This has more of an effect on the suckdown, where there is less offset to the high entrainment, but a weaker fountain is more likely to be entrained into the lift jets or affected by any crosswind.

Widely spaced jet configurations rarely need airframe-mounted devices to control HGI. If ingestion occurs, it is normally directly into the inlet, and often at quite a high height, in which case there is no airframe device that could reduce the HGI. The way to avoid ingestion in these cases is to change the jets such that the fountain does not point at the inlet. This could be accomplished by changing splay or vector, or by subtle alterations to the nozzle to change the spreading rate or the cross-sectional velocity profile.

The JSF F-35 is an example of a widely spaced jet configuration (see Figs. 1.4 and 1.5). The four jets have little similarity to each other, with the roll posts being much smaller and lower thrust than the other two. The lift fan jet has a square cross section, while the main engine jet is circular, but has a severe right angle bend immediately upstream of it. The small roll post jets have some effect on the overall flowfield, but much less than if they had similar strength to the other two. The main fountain develops spanwise between the lift fan and the main engine jet. The angle the fountain makes, and whether there is any ingestion, depends on the relative magnitude of the thrust of the two main jets and the detail of the ground flow patterns from these dissimilar jets.

With no reaction control system, the F-35 changes fore/aft thrust split to trim and control the aircraft in pitch in the jet-borne and semi-jet-borne modes. This affects the strength and inclination of the fountain.

III. Factors Affecting Ingestion

There have been many investigations[11–36] of the factors that affect hot gas ingestion. These factors include wind (or aircraft forward speed), inlet position and inlet shields, as well as aircraft maneuver, pitch and bank angles. Key results from some of these references are discussed in this chapter.

A. Effect of Wind

External wind can have a very strong effect on the HGI characteristics of a configuration. Most work has concentrated on true headwind, with crosswinds not normally being considered. This is because most vertical landing aircraft are difficult to control with crosswinds, and during a vertical landing it is generally straightforward to head into the wind.

A broad generalization is that the closer the lift jet spacing the less effect the wind has, particularly on the fountain. This is directly caused by the strength of the fountain relative to the wind. The Harrier, for instance, sees little effect of wind on HGI until very strong winds, above 30 kn, are encountered. For widely spaced jets winds can have a much bigger impact. A phenomenon seen with a number of research concepts is that relatively high HGI in still-air conditions can reduce to virtually nothing at 10 or 15 kn wind. The fountain in this instance is both weaker and has to travel further to reach the inlet.

The penetration of the ground jet into the wind determines how quickly the jet is rolled up and blown back towards the aircraft. If this distance is high, then the chances are that most of the jet will be entrained back into itself, and little will get back to the inlet. If the distance is very small, then the ground jet might not penetrate as far forward as the inlet, and there will be no ingestion. Between these two extremes lies a critical wind-speed range, the width of which varies with each configuration. Within the critical wind-speed range some of the ground jet will go into the inlets.

The penetration distance depends principally on the value of V_e and the arrangement of the jets, that is, side-by-side front jets will tend to penetrate further than a single equivalent jet, and also on the state of the ground boundary layer. Knowles et al. in Ref. 12 investigated the penetration of the ground jet and found that the height of the nozzle above the ground and the nozzle pressure ratio of the jet made little difference to the penetration at a constant value of V_e. Their twin front jets demonstrated 50% greater penetration than the single jet, while reducing the thickness of the ground boundary-layer reduced penetration. Figure 5.7, from their reference, compares their data with previous investigations for the simple case of a single circular jet impinging perpendicularly on the ground. There is considerable scatter in the data, most of it because, as pointed out in Ref. 25, most of the tests were done in a wind tunnel with a fixed ground board where the ground jet can progress further forward against the lower energy of the freestream boundary layer. For a single jet moving over the ground, the forward extent of the ground jet is given by

$$\frac{x}{D} = \frac{0.5}{V_e} \tag{5.1}$$

If a configuration is found to have a wind-speed range where ingestion is a problem, then operationally it is possible, under some circumstances, to avoid this. For a carrier-borne STOVL it is possible for the ship to alter its speed to

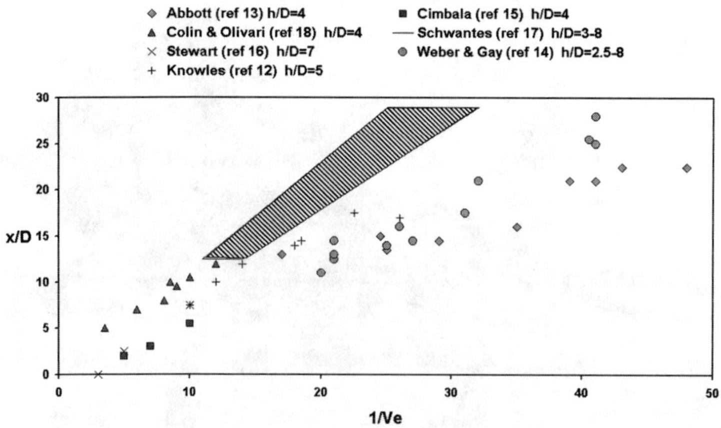

Fig. 5.7 Forward penetration of ground jet.[12]

optimize the wind-over-deck (WOD) speed. For a fixed-base aircraft, where there is room, it is possible to perform a rolling vertical landing (RVL). These are maneuvers where the ground speed is anything up to about 30 kn. Although there is no wing lift to speak of at this speed, the absence of HGI can mean that an RVL gives a higher landing mass than a pure vertical landing (VL). The landing distance at 30 kn is obviously still fairly small.

B. Effect of Inlet Shields

Inlet shields can take the form of either fixed or deployable devices, which are either single purpose or are also used for some other purpose. They divide into solid surfaces and jet shields, both of which have been successfully used in flying aircraft. They are at their most effective when there is a clearly defined flowpath of hot gas close to the aircraft surface, which finds its way into one of the inlets. Inlet shields are mostly ineffective against fountain flows with direct paths into an inlet.

Solid surface shields can be seen on the AV-8B Harrier and are described in Sec. II.A.1. They comprise both fixed strakes (or gun pods) and a retractable cross dam and have a beneficial effect on reducing suckdown close to the ground, as well as reducing gas ingestion. The Yak-39 had a pair of small fixed longitudinal strakes either side of the lift engine inlets (Fig. 5.8). The physical extent of these devices is obvious to see, and their influence on the surrounding flowfield does not extend very far beyond this.

Hall and Rogers in Ref. 11 showed that, where possible, it is important to catch and redirect the fountain flow while it retains enough energy to significantly alter its trajectory. This study used a simple two jet configuration in a longitudinal pod. As shown in Fig. 5.9, the installation of shields at the exit plane of the jets causes a major reduction in the inlet temperature rise, and this reduction appears to be almost independent of the size of the shield. The flowfield and the mechanism by which these reductions are achieved are illustrated in Fig. 5.9. The shields act to deflect the impinging flow laterally away from the body and the inlets.

Fig. 5.8 Russian YAK-38 showing inlet shield installation.

In addition the highly turbulent, laterally deflected upwash field entrains free air from above, causing a downflow at the inlets and on the sides of the body (upper right in Fig. 5.9).

Shields at the plane of the inlets (Fig. 5.10) were, however, relatively ineffective in reducing the inlet temperature rise. The rather sharp corners on the lower side of this body apparently reduced the energy of the flow up the side of the body to the point where it could not be meaningfully deflected laterally to form an entrainment barrier to the sink effect of the inlet.

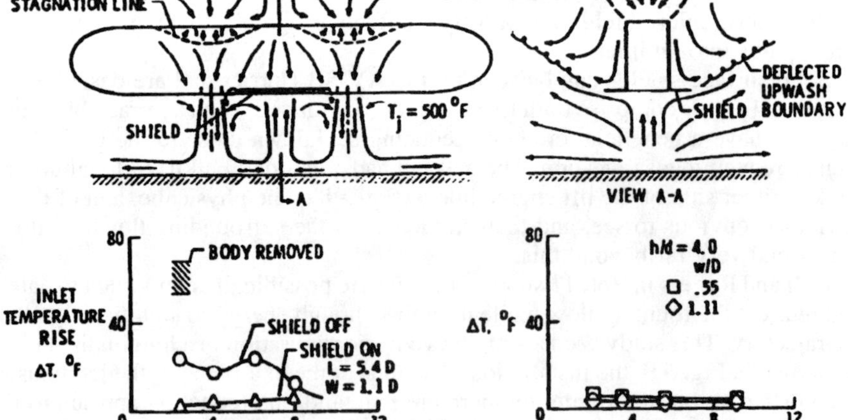

Fig. 5.9 Effect of shields at the jet exit plane in reducing hot-gas ingestion.[11]

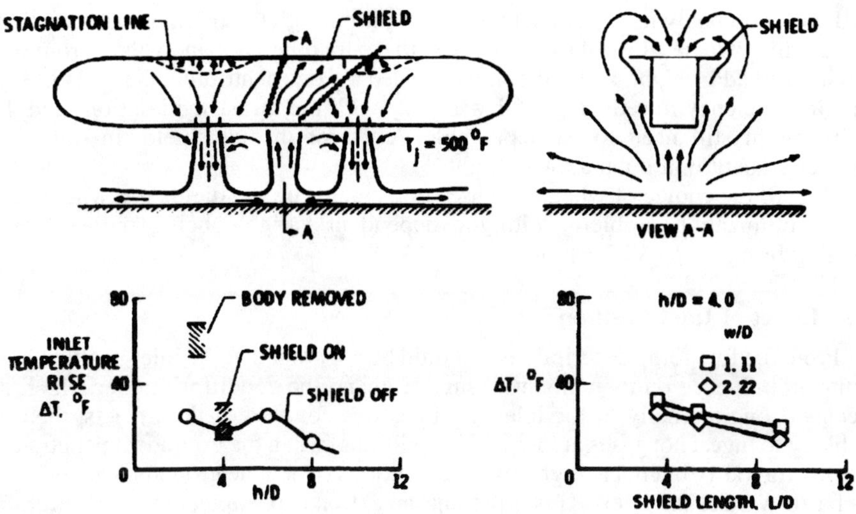

Fig. 5.10 Effect of shields at the inlet plane in reducing hot-gas ingestion.[11]

Jet shields, or screens, can be thought of as acting in much the same way as solid shields, but with the opportunity to extend their influence further from the aircraft surface. The Boeing X-32 JSF demonstrator (Ref. 37) had a transverse jet shield just behind the main aircraft inlet, which can be thought of as being like the Harrier retractable fence. The fountain flow along the fuselage, from the two side-mounted main lift jets, would encounter the jet shield and be deflected down. Boeing had the opportunity to make the jet shield strong enough to prevent this fountain flow from rolling around the shield and being ingested, which occurs with the limited depth of fence on the Harrier. The Boeing proposal for their production JSF aircraft would have added a longitudinal jet shield to the transverse shield. The longitudinal jet shield addressed the ground jet flowfield, which at the plane of the aircraft inlet would have had some upward component from the reinforcement of the side-by-side main lift jets. By disrupting the formation of this reinforcement, at least at low aircraft heights, the upward component of the ground jet can be suppressed. The longitudinal jet shield is a different solution to the problem addressed in Ref. 8 for the Harrier, where it was found that additional strakes forward of the fence reduced the ingestion.

In general jet shields have potential benefits and drawbacks compared to solid shields. On the plus side, 1) they reduce the need for extra structure, whether moving or fixed, which needs to be strong enough, and therefore heavy enough for all conditions where they are deployed; 2) if the solid shields are deployable, then they come with the extra complexity of actuators and the need to maintain signature levels when stowed; and 3) the jet shields obviously have a thrust, which can be modulated to give attitude control. On the negative side, 1) the jet shields need to be pointing downwards to contribute to the total thrust of the aircraft. Even if they are: 2) they need to get their flow from the engine, and this is probably not as efficient as putting it through the main nozzles, so that there is likely to be a loss of total thrust; 3) the extra pipework and valves from the engine will add weight, and if the shield is used to control attitude, then it is likely that its efficiency in reducing HGI will vary with the thrust modulation; and 4) there might still need to be deployable doors for the jet shield, in order to control signature in up and away flight.

Overall the choice of whether to use solid shields or jet shields is a whole aircraft optimization problem, with the dependent variable being the vertical-landing bring-back (VLBB) mass.

C. Effect of Inlet Position

From the foregoing descriptions it should be apparent that the inlet position has quite an influence on the amount of ingestion and the conditions under which it occurs. Longitudinally, if the inlet is a long way forward of the lift jets, then it is likely to ingest hot gases at low wind conditions (when the ground jet penetrates further forward), but not at higher wind conditions (when the ground jet never gets as far forward as the inlet). This is perhaps an extreme example, but for an aircraft with a spanwise fountain it would be best to position the inlets forward of the most forward reach of the fountain, to prevent direct ingestion.

Vertically, it might be thought that underfuselage inlets are most susceptible to ingestion, and this is generally the case. However, the underfuselage inlet is also generally the most easily cured of ingestion, as the flowpath to the inlet is normally simple (direct fountain ingestion, or fountain flow running along the fuselage). Top-mounted inlets are mostly shielded from fountain flows, particularly if above the wing, although there is still a possibility of unforeseen flowpaths. An example would be a fountain flow striking the underside of the wing, traveling forwards (because the wing flaps block the escape path aft), and emerging at the forward wing root. Another flowpath mechanism for HGI of top-mounted inlets is the case of far-field ingestion, particularly in a crosswind. This would be most commonly seen with the aircraft on the ground, possibly preparing for a vertical takeoff.

It should be remembered that jet lift aircraft spend most of their time flying in conventional mode and that allowing jet lift considerations to drive the design to the detriment of conventional performance is a sure way to make a design fail. So although the inlet position will affect HGI, the inlet should really be optimized for up and away.

D. Effect of Maneuver

V/STOL aircraft perform a number of different maneuvers while in their powered lift mode, and each of these presents different HGI characteristics. The most commonly performed, and therefore the most studied, is the vertical landing. In practice, vertical landings are performed at different descent rates, and this can have an effect on the ingestion. The typical HGI trace shows a peak of ingestion at a particular height during the descent. The effect of varying the descent speed is normally to vary the height at which this peak occurs, but not to really alter the magnitude of the peak. A faster descent will shift the peak to a lower height, whereas a slower descent will shift it to a higher height. The magnitude of the shift depends on the length of the flowpath to the inlet. A direct near-field ingestion characteristic will not demonstrate much variation with speed, but an indirect midfield characteristic probably will.

The vertical takeoff (VTO) has not been studied so much in recent years, as the benefits of STO became better understood. All V/STOL aircraft that have flown have still had to demonstrate VTO, and so it does require some investigation. VTO is not as simple as a VL in reverse, for at least two reasons. The first is that VTO is not a constant-velocity maneuver, but an accelerating one. To lift off, there needs to be an excess of thrust over (weight + lift loss). The actual acceleration is rarely constant, as weight is reducing because of fuel burn, lift loss varies with height, and vertical drag (in the downward sense) increases with vertical velocity. From a model test point of view, it is normally simplified to a constant value. Another reason why VTO is different is that in the real case it takes a finite time for the engine to produce enough thrust to lift off. Unless there is some elaborate tie-down/hold-back system, the lift system has to be vectored down while the engine is accelerated, so that a quantity of hot gas is being generated under the aircraft prior to launch. It is important to ensure this gas does not have a flowpath to the inlets at some part of the takeoff. The Harrier has a phenomenon of a "stagnating" VTO, where under adverse circumstances the aircraft can lift off a few feet but no further. This generally occurs at a high weight, where the excess thrust over weight is marginal. With the initial acceleration too slow, the aircraft cannot break through the combination of increased lift loss and HGI that occurs in the 5- to 10-ft region. If the pilot maintains the hover at this height for long enough, rather than aborting the takeoff, he burns off enough fuel to then break through. Good mission planning normally prevents this occurrence.

Creeping vertical landing and takeoffs can be performed on a landing site where the operator is unsure of the integrity of the ground surface. The maneuver is intended to minimize damage to the aircraft should the ground start to break up for any reason. Typically, the aircraft will be moving at about 30 kn, which is not enough to generate any wing lift. On a still day this will be equivalent to a purely vertical maneuver in a 30-kn wind. On a windy day, the equivalent wind strength will be increased, although the effective ground boundary layer will be different. For an aircraft with an HGI problem in still conditions, the creeping VL might be required as an operational procedure.

E. Effect of Changes in Pitch and Bank Angle

Most V/STOL aircraft have been able to be controlled to very tight tolerances of pitch and roll during hover and vertical landing, pilots being very sensitive to attitude changes and the aircraft having high gain control effectors in hovering flight. It would also be impossible for a pilot to maintain an angle of pitch or bank where his jets were not pointing vertically down, as the aircraft would drift away from the landing site.† It is, however, possible to maintain an effective bank angle or pitch angle relative to the ground if the ground is at an angle. This is the case when landing at sea on a moving deck. The worst angles

†In headwind, of course, the jets have to be vectored aft slightly in order to prevent the aircraft drifting backwards. The required vector angle depends on the wind strength and the drag of the aircraft. For the Harrier, pilots use an approximate rule of thumb of 1 deg vector for every 10 kn of wind. This can be achieved by vectoring the nozzles or pitching the nose down by the same amount.

of pitch and roll for small aircraft carriers are in the region of 2 or 3 deg, and their frequency is such that they are quasi-steady for the ground effect portion of a vertical landing.

In these circumstances it is important to know whether the aircraft has an HGI sensitivity to either pitch angle or bank angle. Generally speaking, an effective nose-up pitch is bad, as the angle of impingement strengthens the forward portion of the ground jet and the fountain flow. Bank angle can also be bad, as the lateral shift in the fountain might enable it to flow around any devices placed to capture it.

IV. Estimates of Inlet Temperature Rise

A. Empirical

References 38–41 developed a method of predicting inlet temperature rise for a particular configuration. It divided the flow into a number of zones, such as the freejet zone, impingement zone, upwash zone, etc. The resulting empirical predictions were specific to the configuration they were developed for and dependent on the data which had already been measured. One assessment[42] of the Green and Zanine empirical method showed accurate prediction of inlet temperature rise as a result of far-field recirculation and underprediction of inlet temperature rise as a result of near-field flow. It also identified several deficiencies of the method. The sensitivity of HGI to very small differences in nozzle and jet conditions means that empirical predictions have very limited usefulness.

B. CFD

CFD holds the hope of being able to predict reasonable estimates of inlet temperature rise, given the rapidly developing computing power available to the researcher and engineer. It should, however, be borne in mind that the prediction of circular jet spreading and mixing has been one of the toughest nuts for Reynolds-averaged Navier–Stokes (RANS) codes to crack, and here we are talking about time-averaged values for single jets in free air, let alone time-accurate solutions for multiple jets impinging on the ground.

The anisotropy of turbulent shear stresses within the circular jet is a severe test for the turbulence models used in RANS-type codes, so that it needs another level of sophistication, large-eddy simulation (LES), to start to model them correctly. Unfortunately LES requires another level of computing power from RANS codes, which stretch current (2003) resources to the limit for multiple jet systems. LES is also a time-accurate solution, so that a number of cases probably need to be run in order to obtain an average solution.

It is probably fair to say that RANS codes will accurately model the case where there is going to be no ingestion at all, or where there is going to be severe ingestion. It is very unlikely to come close to accuracy in the case where there is partial ingestion, which might or might not be a problem, or a case where there is occasional ingestion, this being a chaotic system where flow tends to switch around on the margins. The biggest danger is that CFD will falsely reassure the design team that they have no need to worry. There are no published reports of the successful application of CFD to predict HGI at the time of this book. Because there is almost no research activity in this area, it is not known when successful HGI solutions will be demonstrated.

Without being able to predict temperature rise empirically or with CFD, the researchers and design team have to rely on experimental subscale testing.

V. Testing Techniques and Scaling

HGI experimental testing is neither easy nor cheap. The dynamics of the flowfield mean that it is necessary to test dynamically, that is, it is not good enough to test at a series of fixed heights to simulate a vertical landing. The landing maneuver must be performed. For this purpose a dynamic, moving model rig has been developed (Fig. 5.11) at BAE Systems. The unsteady and unstable, indeed chaotic, nature of the flowfield can lead to significant scatter between landings. This means that many repeat tests are needed to accurately characterize mean and peak values of both engine face temperature rise and temperature distortion. Reference 6 shows some examples of the difference seen between dynamic and fixed height testing and also some typical scatter in HGI results between repeats. Figure 5.12 is taken from the reference and shows this behavior. In this plot the symbols show data points collected during a landing, and the solid line is the mean of these. The bars at 0.5, 1.0, and 2.0 m show the variation and mean of data collected during hovers at these heights.

The following sections deal with some of the aspects of subscale model testing for hot-gas ingestion. The premise for this is that we wish to test at smaller than full scale and at temperatures lower than full scale, in order to avoid having to use special materials in our test rig and model. Normal steels can be used up to about 500°C (900°F), and so this is usually the upper limit for any experimental rig.

A. Scaling Considerations[‡]

The forces on the hot jets that create the flowfield around the aircraft come from the nozzle pressure ratio (NPR) and from their buoyancy. The jet dynamic pressure is defined in some references as

$$q = \frac{1}{2}\gamma p M^2 \tag{5.2}$$

in others as

$$q = \frac{F}{2A} \tag{5.3}$$

[‡]This subject is covered in much more depth in Ref. 35.

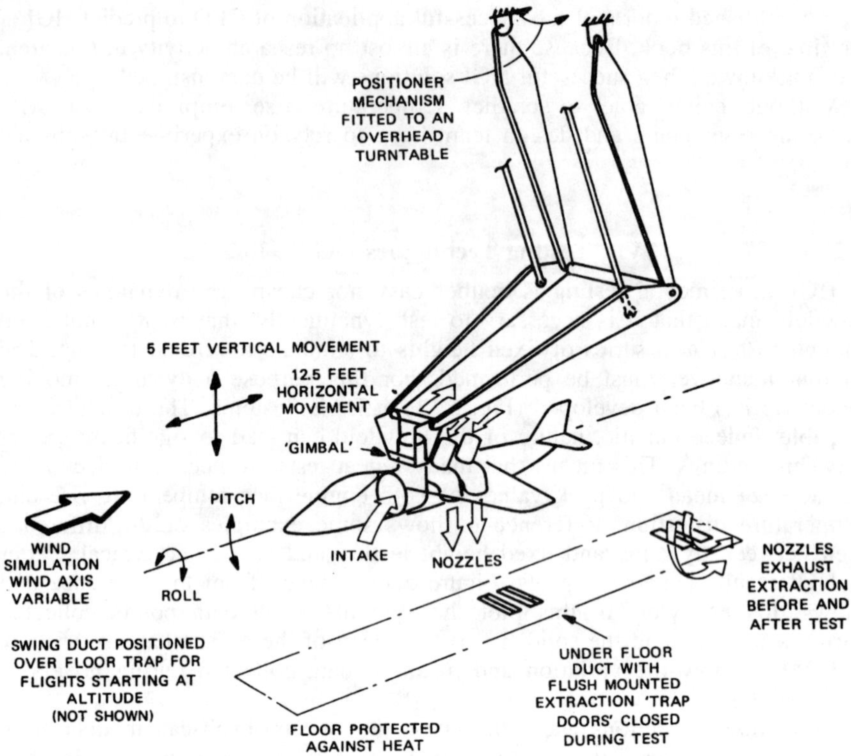

Fig. 5.11 BAE Systems moving model HGI test rig.[36]

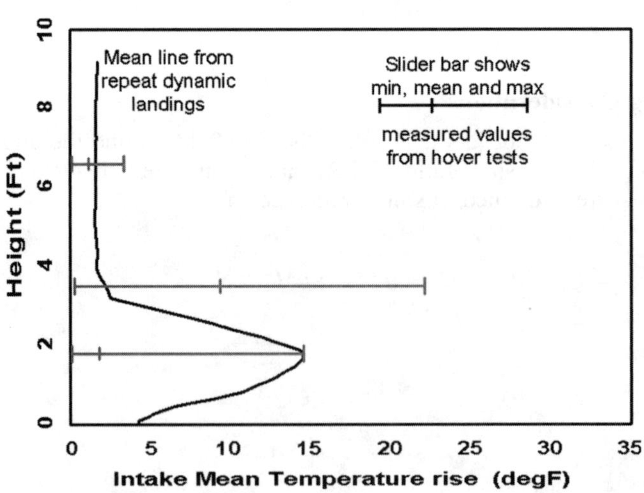

Fig. 5.12 Typical difference between fixed height and moving model test results.[6]

where p is the static pressure, M the Mach number of the jet, F is the thrust, and A the area of the nozzle. Whichever is used makes little ultimate difference provided it is consistent.

The buoyancy of the jets is related to their density, that is, how much lighter they are than the surrounding air. The Cox number, first developed in 1964 at the National Gas Turbine Establishment (NGTE) in the United Kingdom,[31] describes the buoyancy of a jet relative to the dynamic pressure of the jet; thus,

$$\frac{2q_jT_a}{g\theta_jD_j\rho_a}\sqrt{\frac{T_j}{T_a}} \tag{5.4}$$

where g is the gravitational constant, D is the diameter of the jet exit, θ is the excess temperature above ambient, ρ the density, and T the absolute temperature. The subscripts a and j refer to ambient and jet exit conditions, respectively.

It can be seen that to maintain the Cox number for subscale testing, where D is reduced and T_j is normally reduced, it is necessary, because the gravitational constant cannot be changed, to significantly reduce the dynamic pressure of the jets. This would lead to a failure to capture the near nozzle pressure-related effects of supersonic jets, even though the majority of the flowfield would be accurately modeled. It is impossible to correctly match the whole flowfield at less than full scale, so that the best compromise needs to be sought.

In the 1960s and 1970s the engine concepts being considered for STOVL aircraft had relatively low nozzle pressure ratios and high exit temperatures, and it was shown by Cox and Abbott[31] that correct buoyancy scaling was more important in matching the flowfield than dynamic pressure. In recent times, however, the pressure ratios have risen, and the jet temperatures have fallen. This has led to a reappraisal of scaling practice, whereby getting an absolute match of the relative buoyancy is less important than matching the dynamic pressure-driven details of the flowfield. Certain configurations have been found where it is essential to match the NPR of the full aircraft, and others where it is only necessary to ensure the various jet and wind momentums are at the correct ratio. Experimental investigators have to consider these issues with each new configuration they test to find which aspect of the flowfield it is most important to model.

The scaling equations that need to be satisfied were laid down by Cox and Abbott in the early 1960s (Ref. 31), but have been reappraised and tweaked, culminating in Blogg's paper from 2000 (Ref. 35), which shows significant experimental and computational evidence supporting the "root T" scaling method. The scaling laws are noted next. In the subsequent equations the subscripts F, R, I, and W relate to the front jet, rear jet, inlet, and wind, whereas the subscripts fs and ms refer to full scale and model scale.

1) Geometric model to full-scale ratios remain constant:

$$\frac{D_{\mathrm{F,fs}}}{D_{\mathrm{F,ms}}}=\frac{D_{\mathrm{R,fs}}}{D_{\mathrm{R,ms}}}=r \tag{5.5}$$

2) Dynamic head ratios remain constant:

$$\frac{q_{\text{F,fs}}}{q_{\text{F,ms}}} = \frac{q_{\text{R,fs}}}{q_{\text{R,ms}}} = \frac{q_{\text{I,fs}}}{q_{\text{I,ms}}} = \frac{q_{\text{W,fs}}}{q_{\text{W,ms}}} = p \tag{5.6}$$

3) Corrected excess temperature ratios remain constant:

$$\frac{\theta^*_{\text{F,fs}}}{\theta^*_{\text{F,ms}}} = \frac{\theta^*_{\text{R,fs}}}{\theta^*_{\text{R,ms}}} = \frac{\theta_{\text{I,fs}}}{\theta_{\text{I,ms}}} = m^* \tag{5.7}$$

where $\theta^* \equiv \theta(T_a/T_j)^{\frac{1}{2}}$, the so-called root T correction (5.8)

Note also that the inlet temperature rise does not have the root T correction.

4) Corrected buoyancy ratios remain constant:

$$\frac{B^*_{\text{F,fs}}}{B^*_{\text{F,ms}}} = \frac{B^*_{\text{R,fs}}}{B^*_{\text{R,ms}}} \tag{5.9}$$

(Ideally, this ratio would be 1, but see the preceding discussion.)

5) Velocity ratio is

$$\frac{v_{\text{fs}}}{v_{\text{ms}}} = \frac{q^{\frac{1}{2}}_{\text{fs}}}{q^{\frac{1}{2}}_{\text{ms}}} = p^{\frac{1}{2}} \tag{5.10}$$

6) Time ratio is

$$\frac{t_{\text{fs}}}{t_{\text{ms}}} = \left(\frac{D}{v}\right)_{\text{fs}} \bigg/ \left(\frac{D}{v}\right)_{\text{ms}} = \frac{r^{\frac{1}{2}}}{p} \tag{5.11}$$

Of these equations, the velocity ratio and time ratio have interesting fall-outs. The velocity ratio must be applied to the moving model on the dynamic rig (Fig. 5.11). So, if the buoyancy ratio is maintained at 1 (and jet q reduced), the speed of the rig can be reduced to simulate a given aircraft speed. But if it is more important to keep jet q at full-scale values, the rig has to move at full-scale speeds. For the time ratio it can be seen that at full-scale jet q the model time is faster than real time, whereas if buoyancy ratio is maintained then model time slows down, and might be slower than real time. This is important when considering temperature measurements and lag corrections, as discussed in the next section.

B. Temperature Measurements

The whole point of testing for HGI is to measure the temperature rise and temperature distortion at the engine face or aerodynamic interface plane (AIP). Because transient distortions lasting just one engine revolution can trigger a stall, it is important to have very fast response instrumentation. The approach adopted is usually to have very fine wire, open bead thermocouples, normally arranged in a 45- or 48-probe rake at the AIP. Using thermocouples with 0.05-mm (0.002-in) wires, it is possible to have a time constant of the order of 10–15 ms. This is still not really fast enough to capture the extent of transients in the flow, and it is necessary to apply a lag correction to the measured data. The faster the thermocouple time constant, the smaller the correction needs to be. If testing at reduced dynamic pressure, then the time ratio (from the preceding equations) will lead to a smaller lag correction.

The lag correction algorithm used has to make an assumption about the type of temperature change being experienced. Is it a step change or a ramp change, for instance. The different algorithms do not make a big difference to the predicted temperature provided the time constant of the thermocouples is short enough. The simplest algorithm is

$$T_{\mathrm{corr}\,2} = T_{\mathrm{meas2}} + \Delta T_{\mathrm{meas}} \frac{\tau}{\Delta t} \tag{5.12}$$

where corr is the corrected temperature, meas is the measured temperature, subscript 2 is the measurement time, τ is the time constant, and Δt is the time between measurements. This algorithm tends to overshoot with its predictions for very sharp-edged rises by a few percent, and so a better equation is:

$$T_{\mathrm{corr}\,2} = T_{\mathrm{meas1}} + \frac{\Delta T_{\mathrm{meas}}}{(1 - e^{-\Delta t/\tau})} \tag{5.13}$$

where this time it is the first measured temperature that is acted upon.

With such rapid acting instrumentation it is necessary to log the data at a high rate to make sure that all significant transients are captured. It has been found that a logging rate of 250 samples per second is adequate. Given that the temperature of the jets being tested is most likely to be lower than full scale, the measured temperature, after the lag correction has been applied, is scaled up to full scale using Eq. (5.7).

C. Effect of Inlet Flow

The air sucked into the inlet ought to be scaled correctly to the jet flows in the model, that is, the momentum ratio of the model inlet flow to full scale should match that of the jets. It has been found, however, that reduced inlet flows do not necessarily change the measured HGI characteristics. The reduction can be as high as 50% before a difference is seen (Fig. 5.13). This is quite useful, as the packaging constraints of HGI models often make it difficult to achieve the desired level of suction. The reason for the inlet flow not being critical is that inlets do not really drive the flowfield, but merely sample what is presented to them.

This is an outcome of the suction strength being proportional to $1/r^2$, and in a practical sense can be thought of like a vacuum cleaner nozzle. The suction gradient is very high, there being very strong suction at the plane of the nozzle, but very light suction only a short distance away.

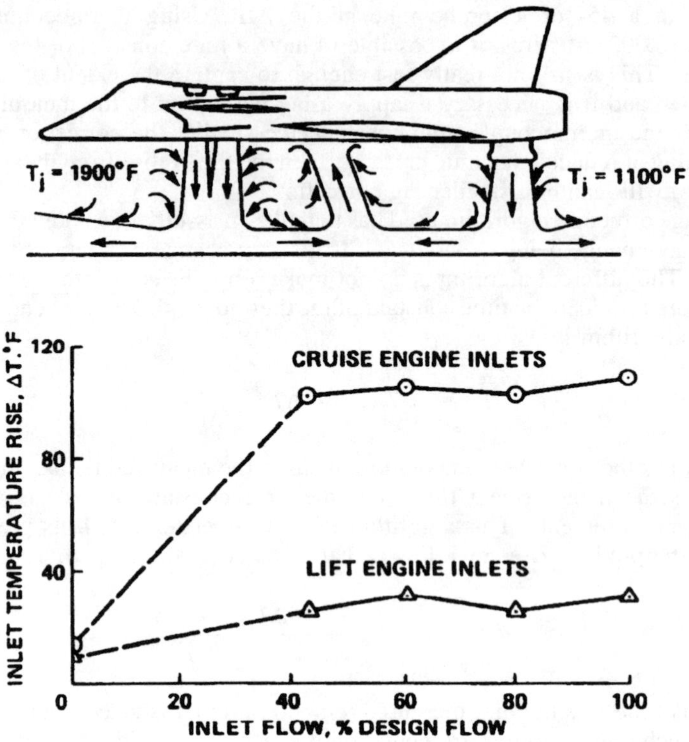

Fig. 5.13 Effect of inlet flow on HGI.[14]

D. Jet Modeling

The most important aspect for any HGI modeling, whether experimental or CFD, is to correctly characterize the jets.[37,43,44] This is very difficult for early conceptual work where there is very little known about the real nozzles and jets, but the researcher must always use the best information available and always do sufficient calibration work to know what has been tested. Usually this means simply matching the geometry of the full-scale hardware, but if there are full-scale data available for plume shapes, decay rates, ground jet profiles, etc., then they need to be matched.

E. Effect of Scatter

Some combinations of configuration and test conditions will give very repeatable data, with little scatter between repeats. Other combinations show very large scatter between successive repeats. Often this is because the ingestion is sensitive to the condition of the fountain flow, which is very unstable and exhibits chaotic behavior.

Usually it is found that the average characteristics measured from a series of repeats will match extremely well with a similar series of repeats taken at a different time, even if the scatter within each set is as high as 100% of the mean of the measured peak value.

Low scatter configurations normally need about five repeats to exhibit true mean values, whereas high scatter configurations can need 10 or more repeats to demonstrate an average characteristic.

For temperature distortion, as discussed earlier, it is more important to know the likely peak value, than the mean value at a height, as this will drive the compatibility considerations for the engine. If there is high scatter in the peak value, then it is likely that over the course of a large number of landings a much higher distortion will occur. The use of extreme value statistics enables predictions to be made of the worst distortion likely to be seen during, say, 10,000 landings, from a set of 10 or so repeats. The technique proposed by Beasley[45] of Rolls–Royce is the one commonly used in industry. This is based on the work of Weibull, which is commonly used in reliability engineering and life data analysis.

Beasley took the peak distortion value experienced during each landing from a set of repeats and plotted these values on the special Weibull graph paper. If the data form a straight line, then it can be said that they come from a valid sample sharing the same ingestion phenomenon, and the data can be extrapolated. If the data form a curve or perhaps appear to have two straight lines with a kink, then there is more than one phenomenon occurring, and the analysis becomes much more difficult. Experience in industry, Ref. 6 for example, shows that under most conditions the standard Weibull analysis works well. The two or more phenomena cases tend to occur at a switchover point. An example would be where fountain flow is directly ingested at low headwind speeds, but not at higher speeds. There will be a band of wind speed where sometimes ingestion occurs and sometimes it does not. If tests are conducted in this wind speed, then the Weibull analysis becomes difficult.

VI. Techniques to Reduce HGI

The preceding sections of this chapter have shown that HGI is caused by fountain flows and ground jets being in the wrong place at the wrong time. The techniques needed to reduce or eliminate HGI must therefore be concerned with either controlling these flows or in preventing them occurring in the first place. Control really means the use of inlet shields, which, as already seen, take the form of either solid shields or jet screens. In both cases there is a penalty to pay in terms of weight, performance, and complexity, and hence reliability and maintainability. Prevention usually takes the form of altering jet splay and vector angles or plume profiles, such that the fountain either does not form or is pointing in a different direction to the one that causes ingestion.

A. Closely Spaced Jets

For configurations with closely spaced jets, it is very difficult to totally prevent fountain formation without having a significant cosine thrust loss. For the Harrier we saw that by having greater splay on the rear nozzles than on the front the fountain is angled slightly aft, reducing the flow along the fuselage into the inlets. If the jets were all vertical, the fountain flow would be much stronger, and more flow would reach the inlets. The splay is not enough in itself to eliminate ingestion (particularly with the gun pods installed), and so inlet shields are also required.

When trying to eliminate HGI on a closely spaced jet configuration, the engineer must look at the effects of splay before designing inlet shields. He should also be careful that a change which appears beneficial at one condition works at most other conditions where the aircraft will be asked to perform.

B. Widely Spaced Jets

For configurations with widely spaced lift jets, it has been found that altering jet splay and plume profile are the most effective tools, while inlet shields have limited use.

A British Aerospace/Rolls–Royce research design from the mid-1980s had, among other layouts, two closely spaced lift jets at the back of the aircraft and a single jet at the front. This configuration suffered serious ingestion, which could not be cured by any combination of longitudinal or lateral shields. It was discovered[5] that outward splay on the rear nozzles could completely eliminate the ingestion, as seen in Fig. 5.14. This figure shows that, for descending flight, with zero splay the fountain between the rear jets is not formed until the aircraft has descended to a height of about 3 m, when there is a huge increase in ingestion as this longitudinal fountain flow pushes the main lateral fountain forward to the inlets. Increasing splay up to 6 deg means that the fountain is formed at about 7 m, but by being weaker the peak ingestion is lower. At 10 deg of splay, the fountain between the rear jets is weak enough that there is virtually no ingestion at all.

The McDonnell Douglas/British Aerospace Joint Advanced Strike Technologies (JAST) design, Fig. 5.15, (Ref. 46) had a similar lift jet configuration (but with a convergent nozzle operating at a supercritical NPR leading to an underexpanded exit flow). In this instance it was found that rear nozzle splay, although being generally beneficial, did not completely eliminate ingestion. Flow visualization revealed a flowpath for fountain flow from the underside of

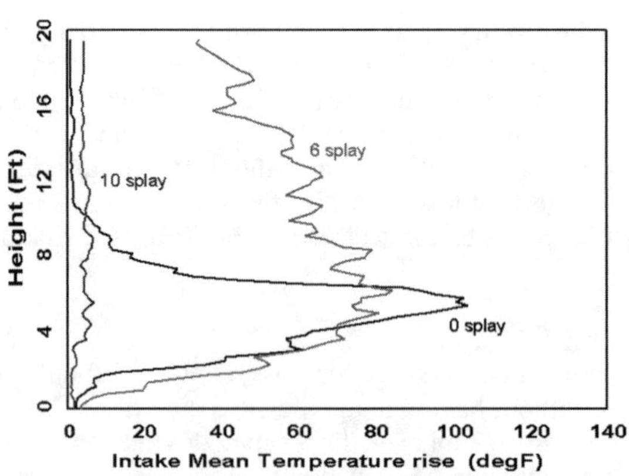

Fig. 5.14 Effects of splay on HGI.

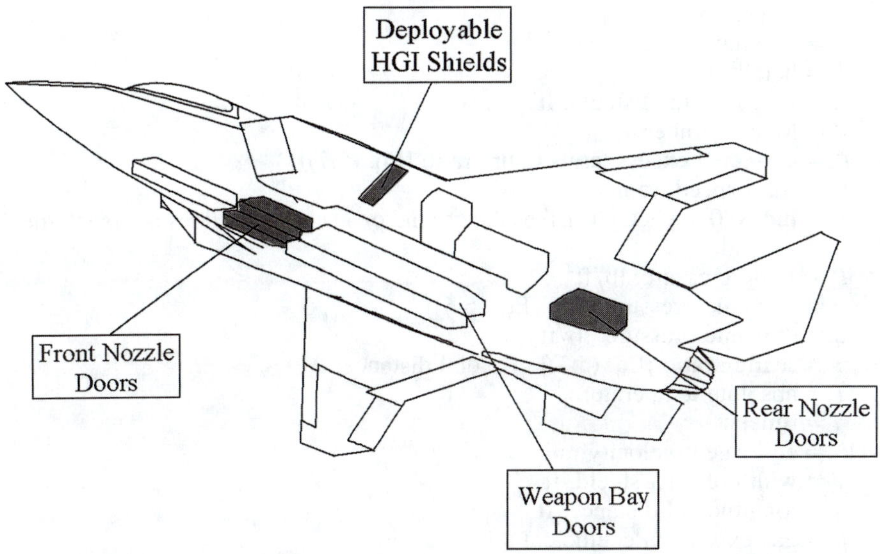

Fig. 5.15 JAST configuration.

the wing, along the angled side of the inlet cowl and into the inlet. A shield had to be placed here to block this path.

Even with only a single jet at the front and the back, it is possible to alter the fountain angle without changing the thrust split or the overall vector (which must equal 90 deg, of course). This is done by altering the relative strengths of the ground jet at front and back, which meet to form the fountain. This can only be done for widely spaced jets where there is enough distance for the ground jet characteristics to be developed. Consider an elliptical-shaped jet plume. This might be thought to be expanding along the major axis and making the flow stronger in the major axis. When this plume impinges on the ground, the flow turns through 90 deg relative to the isobars in the freejet plume, and thus the ground jet is stronger in the minor axis. The axes switch. So to weaken the forward strength of the ground jet from the rear nozzle, it is necessary to give the freejet plume a fore/aft major axis. Alternatively, to strengthen the aft strength of the ground jet from the front nozzle it is necessary to increase the lateral spreading rate (or splay) of the front jet, giving it a lateral major axis. These effects can be achieved by making detail changes to nozzle hardware, while hopefully maintaining performance in the up and away mode.

Nomenclature

A = jet exit area, ft^2
B^* = buoyancy parameter, $(V_j^2/\beta g \theta_j D_j)(T_a/T_j)^n$
d, D = nozzle diameter: actual value for circular nozzles; or effective diameter for noncircular nozzles, ft

F = thrust, lb
g = gravitational constant
h = height, ft
L = length of inlet shield, ft
M = Mach number
m^* = corrected excess temperature ratio [Eq. (5.7)]
N = rotor speed, rpm
n = index: 0.5 when jet strikes the ground or 1.0 if the jet does not reach the ground
P = static pressure, $\mathrm{lb/ft^2}$
p = dynamic pressure ratio [Eq. (5.6)]
q = dynamic pressure, $\mathrm{lb/ft^2}$
r = scaling ratio [Eq. (5.5)] or radial distance, ft
T = absolute temperature, °R
t = time, s
V_e = equivalent velocity ratio, $= (q_j/q_a)^{0.5}$
W = width of inlet shield, ft
x = longitudinal distance, ft
β = gas expansion coefficient
γ = ratio of specific heats
Δt = time between measurements, s
ΔT = inlet temperature rise
ρ = density, $\mathrm{lb \cdot s^2/ft^4}$
σ = air density
θ = excess temperature above ambient, °F
τ = thermocouple time constant

Subscripts

a = ambient
corr = corrected
F = front jet
fs = full scale
j = jet exit
meas = measured
ms = model scale
R = rear jet

References

[1]Penrose, C. J., and Smith, A. P., "Hot Gas Ingestion Control from an Engine Viewpoint," SAE P-306, *International Powered Lift Conference Proceedings*, Society of Automotive Engineers, Pennsylvania, PA, March 1997, pp. 83–93.

[2]Milford, C. M., "Hot Gas Recirculation in V/STOL," SAE P-203, *Proceedings of the International Powered Lift Conference*, Society of Automotive Engineers, Pennsylvania, PA, 1987, pp. 19–30.

[3]Knott, P. G., and Milford, C. M., "Configuration Effects on the Intake of Hot Gas into the Engine Intake," *Royal Aeronautical Society International Powered Lift Conference*, Royal Aeronautical Society, London, UK, 1990, pp. III.3.1–III.3.13.

[4]Johns. A. L., Neiner, G. H., Bencic, T. J., Flood, J. D., Amuedo, K. C., Strock, T. W., and Willimas, B. R., "Hot Gas Ingestion Characteristics and Flow Visualization of a

Vectored Thrust STOVL Concept," *Royal Aeronautical Society International Powered Lift Conference*, Royal Aeronautical Society, London, UK, 1990, pp. III.4.1–III.4.19.

[5]Williams, D. D., "Hot-Gas Reingestion, Engine Response Considerations," *Royal Aeronautical Society International Powered Lift Conference*, Royal Aeronautical Society, London, UK, 1990, pp. III.5.1–III.5.13.

[6]Curtis, P., and Penrose, C. J., "Recent Exhaust Gas Ingestion Experiments on a Generic ASTOVL Aircraft," AIAA Paper 93-4890, American Institute of Aeronautics and Astronautics, Reston, VA, Dec. 1993.

[7]McLemore, H. C., "Considerations of Hot-Gas Ingestion for Jet V/STOL Aircraft," NASA SP-116, *Conference on V/STOL and STOL Aircraft*, 1966, pp. 191–204.

[8]Moss, G. M., and Penrose, C. J., "Development of Modified Lift Improvement Devices for the AV-8B Harrier," AIAA Paper 95-0532, American Institute of Aeronautics and Astronautics, Reston, VA, Jan. 1995.

[9]Phillips, H. I., and Sharland, M. S., "Harrier International Programme," *Royal Aeronautical Society International Powered Lift Conference*, Royal Aeronautical Society, London, UK, 1990, pp. II.2.1–II.2.11.

[10]Curtis, P., "The Influence of Ground Boundary Layer on Hot Gas Ingestion Characteristics," *International Powered Lift Conference*, AIAA Paper 2002-5983, American Institute of Aeronautics and Astronautics, Reston, VA, Nov. 2002.

[11]Hall, G. R., "Model Tests of Concepts to Reduce Hot Gas Ingestion in VTOL Lift Engines," NASA CR-1863, July 1971.

[12]Knowles, K., Bray, D., Bailey, P. J., and Curtis, P., "Impinging Jets in Cross-Flow," *Royal Aeronautical Society International Powered Lift Conference*, Royal Aeronautical Society, London, UK, 1990, pp. Res2.1–Res2.14.

[13]Abbott, W. A., "Studies of Flowfields Created by Vertical and Inclined Jets When Stationary or Moving over a Horizontal Surface," Aeronautical Research Council, C.P. No. 911, London, 1967.

[14]Weber, H. A., and Gay, A., "VTOL Reingestion Model Testing of Fountain Control and Wind Effects," AIAA Paper 75-1217, American Institute of Aeronautics and Astronautics, Reston, VA, 1975.

[15]Cimbala, J. M., Stinebring, D. R., Treaster, A. L., and Billet, M. L., "Experimental Investigation of a Jet Impinging on a Ground Plane in the Presence of a Cross-Flow," Applied Research Lab., Pennsylvania State Univ., Contractors Report to NADC, State College, PA, Feb. 1987.

[16]Stewart, V. R., Kuhn, R. E., and Walters, M. M., "Characteristics of the Ground Vortex Developed by Various V/STOL Jets at Forward Speed," AIAA Paper 83-2494, American Institute of Aeronautics and Astronautics, Reston, VA, Oct. 1983.

[17]Shwantes, E., "The Recirculating Flow Field of a VTOL Lifting Engine," NASA TT F-14912, June 1973.

[18]Colin, P. E., and Olivari, D., "The Impingement of a Circular Jet Normal to a Flat Surface With and Without Cross Flow," von Karman Inst. for Fluid Dynamics, Rep. AD688953, Rhode St.Genesse, Belgium. Jan. 1969.

[19]McKinzie, B. A., et al., "P.1127 (XV-6A) V/STOL Handling Qualities Evaluation," Air Force Flight Test Center, Edwards Air Force Base FTC-TR-68-10, Aug. 1968.

[20]McLemore, H. C., and Smith, C. C., Jr., "Hot Gas Ingestion of Large Scale Jet VTOL Fighter Type Models," NASA TN D-4609, 1968.

[21]Dudley, M., Falarski, M., Pisano, A., and Hill, W., "Ground Effect Hover Characteristics of a Large Scale Twin Tilt-Nacelle V/STOL Model," AIAA Paper 81-2609, American Institute of Aeronautics and Astronautics, Reston, VA, Dec. 1981.

[22]Gittner, U., et al., "Interaction Between Airframe Powerplant Integration and Hot Gas Ingestion for Jet-Lift V/STOL Transport Aircraft," AGARD, 31st Flight Mechanics Panel Meeting on Integration of Propulsion Systems in Airframe, Sept. 1967.

[23]Holzhauser, C. A., Morello, S. A., Innis, R. C., Patton, J. M., Jr., "A Flight Evaluation of a VTOL Jet Transport Under Visual and Simulated Instrument Conditions," NASA TN D-6754, March 1972.

[24]Wood, D. W., Jr., "The Do-31 V/STOL Jet Transport," *Society of Experimental Test Pilots Eleventh Symposium Proceedings*, Society of Experimental Test Pilots, California, 1967.

[25]Kuhn, R. E., "Design Concepts for Minimizing Hot Gas Ingestion on V/STOL Aircraft," AIAA Paper 81-1624, Aug. 1981, also *Journal of Aircraft*, Vol.19, No.10, 1982.

[26]Berenrt, R., and Roekmojoto, R., "VAK Dokumentetion Rezirkulation und Strahlinduktion (Recirculation and Jet Induced Effects)," VFW-Fokker, 1973; (also, NASA TT F-15932) (translation).

[27]Kirk, J. V., and Barrack, J. P., "Exhaust Gas Reingestion Studies on Lift Engine VTOL Fighter Configurations," *Prediction Methods for V/STOL Propulsion Aerodynamics*, Naval Air Systems Command, Patuxent River, MD, 1975, pp. 304–333.

[28]Tolhurst, W. H., and Kelly, M. W., "Characteristics of Two Large Scale Jet Lift Propulsion Systems," NASA SP-116, *Conference on V/STOL and STOL Aircraft*, April 1966, pp. 205–228.

[29]McLemore, H. C., Smith, C. C., Jr., and Hemeter, P. G., "Generalized Hot Gas Ingestion of Large Scale Jet VTOL Fighter Type Models," NASA TN D-5581, Jan. 1970.

[30]Kirk, J. V., and Barrack, J. P., "Reingestion Characteristics and Inlet Flow Distortion of V/STOL Lift Engine Fighter Configurations," NASA TN D-7014, Dec. 1970.

[31]Cox, M., and Abbott, W. A., "Studies of Flowfields Created by Single Vertical Jets Directed Downwards Upon a Horizontal Surface," NGTE, Memo M.90, Royal Aircraft Establishment, UK, Oct. 1964.

[32]Kuhn, R. E., "Hot Gas Ingestion and the Speed Needed to Avoid Ingestion for Transport Type STO/VL and STOL Configurations," AIAA Paper 84-2530, Nov. 1984.

[33]Stewart, V. R., and Kuhn, R. E., "A Method for Estimating the Propulsion Induced Aerodynamic Characteristics of STOL Aircraft in Ground Effect," NADC 80226-60, Naval Air Development Center, Pennsylvania, PA, Aug. 1983.

[34]Wardwell, D. A., Bellavia, D. C., Corsiglia, V. R., and Kuhn, R. E., "Dynamic Response of Induced Pressures, Suckdown and Temperature for Two Tandem Jet STOVL Configurations," NASA TM-103934, July 1992.

[35]Blogg, K., "Model to Full-Scale Temperature Scaling for STOVL Aircraft Model Testing," *International Powered Lift Conference Proceedings*, American Helicopter Society, Virginia, 2000.

[36]Bore, C. L., "Ground Based Testing Without Wind Tunnels," NATO Advisory Group for Aeronautics Research and Development, AGARD-R-710, *Special Course on V/STOL Aerodynamics*, France, 1984, pp. 10,1–10,6.

[37]Margason, R. J., Arledge, T., Wardwell, D. A., Hange, C. E., and Naumowicz, T., "Influence of Jet Efflux Characteristics on STOVL Aircraft Propulsion-Induced and Ground Effects," *Royal Aeronautical Society Proceedings*, Royal Aeronautical Society, London, UK, 1998, pp. 25-1–25-3.

[38]Green, K. A., and Zanine, J. J., "Computerized Method for an Estimate of Hot Gas Reingestion for a VTOL Aircraft at the Conceptual Design Stage," NADC 78256-60, Naval Air Development Center, Pennsylvania, PA, Aug. 1979.

[39]Green, K. A., and Zanine, J. J., "REINGEST—Users Manual," NADC 80225-60, Naval Air Development Center, Pennsylvania, PA, Aug. 1979.

[40]Green, K. A., and Zanine, J. J., "Estimation of Hot Gas Reingestion for a VTOL Aircraft at the Conceptual Design Stage," *V/STOL: An Update and Overview*, Society of Automotive Engineers, Pennsylvania, PA, 1984, pp. 51–65.

[41]Gray, L., and Kiesielowski, E., "Practical Engineering Methods for Predicting Hot Gas Reingestion Characteristics of V/STOL Aircraft Jet Lift Engines," NASA CR-111845, February 1971.

[42]Mourtos, N. J., and Margason, R. J., "Evaluation of a Prediction Method for V/STOL Aircraft Hot Gas Ingestion," NASA TM-103828, Jan. 1991.

[43]Cook, R., Curtis, P., and Fenton, P., "State of the Art in Sub-Scale STOVL Hot Gas Ingestion Wind Tunnel Test Techniques," Society of Automotive Engineers, Paper 2005-01-3158, Oct. 2005.

[44]Curtis, P., "A Review of the Status of Ground Effect/Environment Technologies," *International Powered Lift Conference*, AIAA Paper 2002-5985, American Institute of Aeronautics and Astronautics, Reston, Virginia, Nov. 2002.

[45]Beasley, R., "Unsteady Aspects of Hot Gas Reingestion and Statistical Analysis," *AGARD Conference Proceedings*, CP-534, NATO Advisory Group for Aeronautics Research and Development, France, 1993.

[46]Illston, L. W., and Bradley, P. J., "Environmental Factors Related to JSF STOVL Design," *International Powered Lift Conference Proceedings*, Society of Automotive Engineers, Pennsylvania, PA, 1997, pp. 377–389.

[47]Curtis, P., Xu, L., and Ford, P., "The Design, Development and Testing of a Turbine Powered Simulator for Hot Gas Ingestion Testing," Society of Automotive Engineers, Paper 2000-01-3159, *International Powered Lift Conference and Exhibit*, Grapevine, Texas, Oct. 2005.

[48]Reed, L. H., "Recent Developments at the Shoeburyness STOVL Test Facility," *Royal Aeronautical Society International Powered Lift Conference*, Royal Aeronautics Society, London, UK, Aug. 1990, pp. III.2.1–III.2.8.

Chapter 6

Ground Environment

I. Overview

ALTHOUGH JET- and fan-powered V/STOL aircraft, like the Harrier or the F-35B JSF, can be operated independently from conventional runways, they cannot be operated from the full range of austere takeoff and landing sites that helicopters can. Potential problems that need to be considered include 1) the ability of the aircraft structure to survive the fountain temperatures and acoustics environment of vertical operations near the ground; 2) the ability of personnel to survive and work in the outwash flowfield close to an operating aircraft, because of either ground jet velocity, ground jet temperature, or acoustics; 3) spray generation from sources of water near the landing pad; and 4) damage to the ground by the impinging lift jet. Not only could this lead to unserviceability of the landing pad, but debris could give rise to a) foreign object damage (FOD) to the aircraft caused by particles and objects being lifted into the fountain flows (This damage could be to the fuselage lower surface or to the engines. Fast jet aircraft cannot afford to have particle separators on their inlets like helicopters) and b) FOD to surrounding equipment and personnel from particles and objects carried in the ground jets.

Each of these aspects will be dealt with in this chapter following a brief survey of the physics of impinging jet flows.

II. Impinging Jet Flows

Jet-powered V/STOL aircraft require high specific thrust jets in order to achieve their up-and-away performance. Such jets are not ideal for hovering flight from a number of perspectives, such as their high disc loading (leading to poor hover efficiency) and their impingement properties. The F-35B JSF uses the shaft-driven lift fan, which extracts power from the engine and reduces the disc loading of the lift jets down to the level of the Harrier. Even here, the nozzle pressure ratio (NPR) of both the lift fan nozzle and the aft nozzle give a marginally supercritical flow, that is, the nozzles are choked, with a Mach number of one at the throat.

In a subcritical nozzle flow the jet efflux downstream of the nozzle exit is characterized as depicted in Fig. 2.1. There is a conical region of constant temperature and velocity immediately downstream of the nozzle, surrounded and followed by a turbulent mixing region increasing in diameter and mass as the jet velocity decays with distance downstream.

At supercritical pressure ratios the flow from a convergent nozzle is not fully expanded at the exit from the nozzle. Figure 6.1 shows the variation of the Mach number on the centerline, with distance downstream, at a nozzle pressure ratio of 3.5. This variation is a result of the formation of the well-known shock diamonds as the flow expands and recompresses as it progresses downstream from the nozzle. The convergent-divergent (con-di) nozzle, when operating at its design point, fully expands the flow before the exit, so that this variation in Mach number is not seen. To ensure that the nozzle is at the fully expanded design point throughout the flight envelope, and the throat has the correct area for the engine's needs, requires independent control of both the throat and the exit areas. In practice, the weight and complexity of independent control outweigh the performance benefit, and most nozzle designs have a fixed linkage, and hence fixed area ratio, between the throat and the exit. The consequence of this is that the con-di nozzle will only be truly on design for a limited portion of the flight envelope, chosen to be where the overall benefit is greatest, usually at high transonic speed. For the V/STOL aircraft using a combined lift/cruise nozzle, like the F-35B JSF, the flow will normally be overexpanded at the exit when in V/STOL mode, that is, the Mach number is too high for the NPR. The result of this is again to have an oscillation of the centerline Mach number

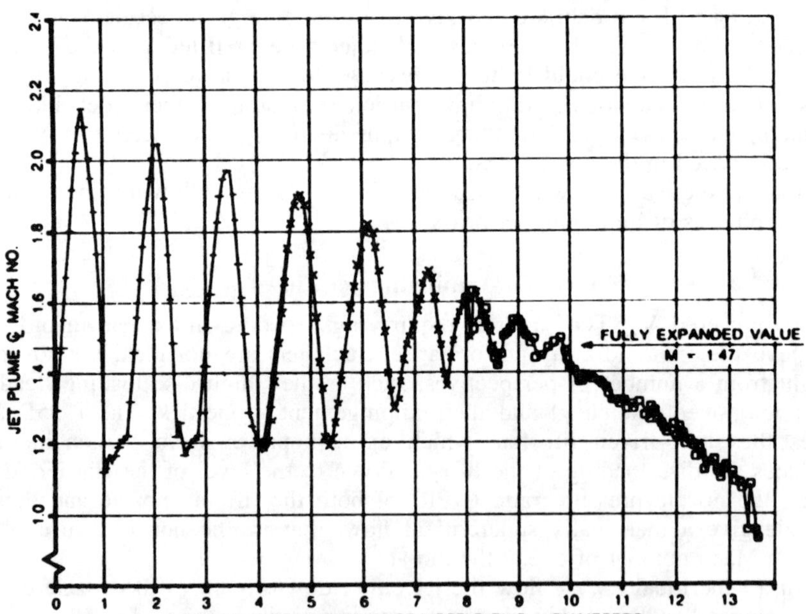

Fig. 6.1 Jet plume centerline Mach-number variation with distance downstream; NPR = 3.5.[1]

as the flow compresses then expands, albeit with a rather stronger first shock than for the underexpanded case.

As stated in Ref. 1, "This shock cell structure inhibits the turbulent mixing process, which therefore takes place more slowly than in the sub-critical jet. The edge mixing process gradually erodes the amplitudes of the shock cell until they damp out at or close to the ideal fully expanded Mach number. This process has taken a downstream distance of about 10 diameters in the example shown, nearly twice the length of the potential core in the subsonic jet."

These changes in the jet characteristics have two effects on the ground environment situation. The aircraft will be subject to higher jet pressure and velocity fluctuations, and the ground will be subjected to higher dynamic pressures and temperatures. Figure 6.2 is a sketch of the structure of an impinging supersonic jet where the underexpansion of the jet is sufficient to produce a Mach disc in the jet. As indicated in Ref. 1, this flowfield gives rise to 1) a total pressure loss through the shocks, 2) an unsteady separated flow region, giving rise to twin pressure peaks, on the ground, and 3) a surface air temperature distribution similar to the subsonic case, that is, no total temperature losses.

Under different conditions, where there is no Mach disc, the flow still has to decelerate to very low speed in the impingement region (stagnation at the impingement point), and a standoff shock exists. The sonic line on Fig. 6.2 would close over the impingement region, like the stagnation bubble beneath it. Either way, the forced deceleration of the flow at impingement is accomplished through a normal shock. Total pressure loss through a normal shock increases with increasing Mach number, and this can be seen in Fig. 6.3, which shows the total pressure ratio at the impingement point, for a convergent nozzle, plotted against the nozzle exit total pressure ratio.

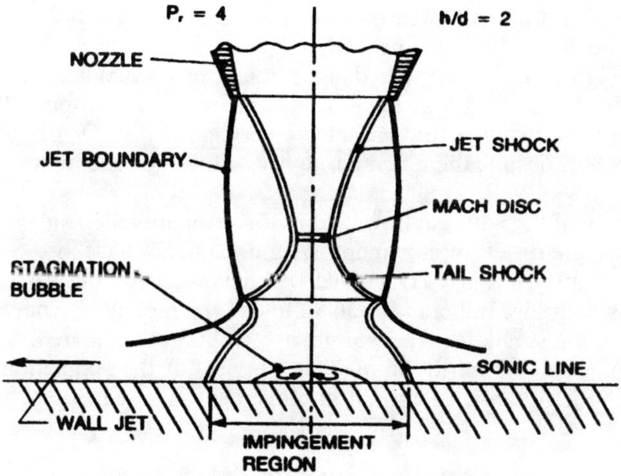

Fig. 6.2 Structure of the impinging supersonic jet.

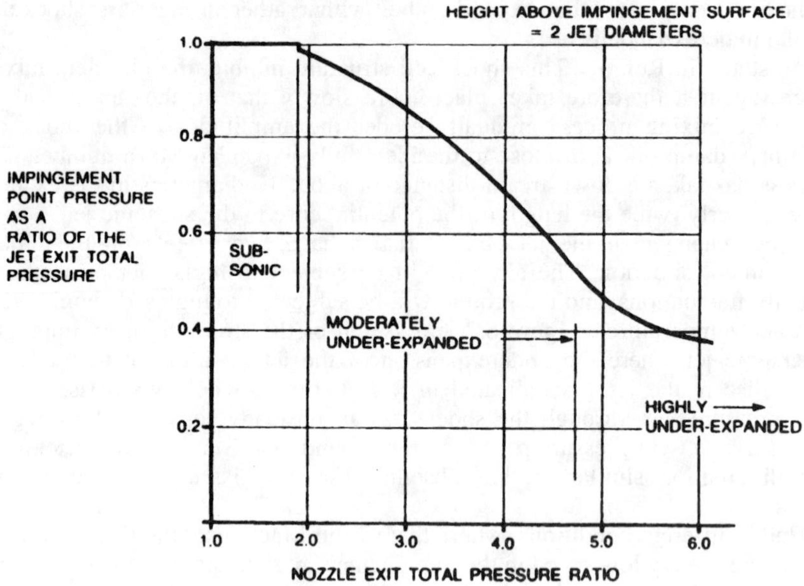

Fig. 6.3 Impingement pressure losses for the flow from a convergent nozzle.[2]

III. Outwash Flows

Outwash flows from the wall jets lead to areas where the strength of the jet, or its temperature, or its acoustic properties would be dangerous to people or objects. The danger could include being blown off your feet, being singed, suffering irreparable hearing damage, or setting up chest cavity resonance, leading to organ damage. Outside these areas there will be a further area where, although it is not dangerous, it is still unreasonably difficult to perform tasks that would be needed on the flight line. Studies related to these problems are presented in Refs. 3–8. Figure 6.4 shows the danger zones recommended for the AV-8B Harrier (Refs. 3 and 4). These have been derived mostly empirically and have not actually been changed throughout the service of the XV-6A Kestrel up to the latest AV-8B, despite there being roughly a 50% increase in thrust. Acoustics are covered in Sec. VI.

The strength of the outwash flow, and its directionality, depend on a number of factors. The angle of jet impingement is significant, so that those configurations with vector or splay in the VTOL mode will have stronger outwash in the direction the jet is pointing. Rubel in Ref. 6 validated the model of Schach to estimate this effect and the strength of the wall jet at any azimuthal angle from the impingement point. Away from the immediate vicinity of the stagnation region, the strength of the jet relative to normal impingement is given by

$$\frac{\rho U_\phi^2}{\rho U_{90}^2} = \frac{\sin^3\theta}{(1-\cos\theta\cos\phi)^2} \tag{6.1}$$

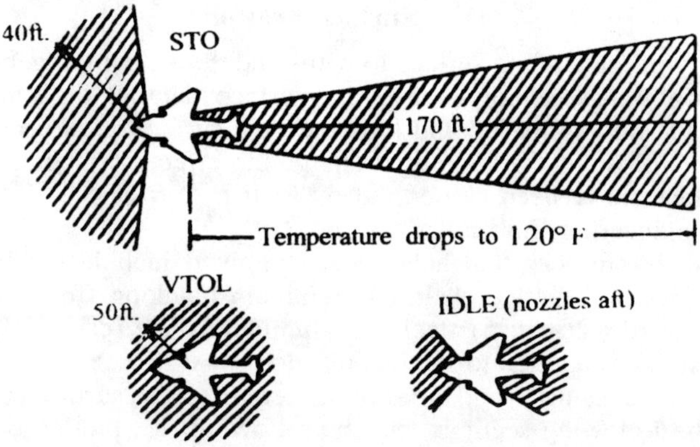

Fig. 6.4 AV-8B Harrier danger areas.[3,4]

where θ is the inclination angle of the jet to the ground and ϕ is the azimuthal angle from the stagnation point, where 0 is downstream in the plane of symmetry. The stagnation point is upstream of the impingement point of the jet centerline by the distance $r \cos \theta$, where r is the radius of the freejet just prior to impingement. Thus, for a 10 deg splay angle (or 80 deg impingement angle), the outwash will be 1.4 times stronger than for a normal impingement, and the inwash will be 0.69 times as strong. Even for 5 deg splay the changes in strength need to be taken into account, the outwash being nearly 20% stronger.

For a given thrust, it is found that NPR has no effect away from the immediate vicinity of the nozzle, that is, in the far field. For this reason the U.S. Marine Corps has had more outwash problems from its CH-53 helicopters, with their very low NPR but high thrust, than from its AV-8Bs, with their much higher NPR.

Reinforced wall jets occur along stagnation lines between two impinging jets. The fan shape of the fountain, discussed in Chapter 2 and illustrated in Fig. 2.11 is seen as a reinforcement along the stagnation line in the far field. The reinforcement region has higher velocity and significant thickening of the wall jet, compared to that from a single jet having the combined thrust of the two constituent jets. Depending on nozzle spacing, splay, height above ground, etc. the reinforcement can more than double the peak velocity (Ref. 7).

The thickening of the jet is also significant. Studies show that peak outwash velocities from VTOL aircraft, around the edge of the keep-out zone, occur at about knee height, with a rapid reduction above this. Reinforced wall jets can maintain their peak velocities up to 8 ft or so. Reference 7 has more detail on all of this.

Although the Harrier, with its outward-splayed nozzles and four lines of reinforcement, has a circular VTOL danger zone, the F-35B JSF will be different. The two main lift jets will dominate, and the roll jets will have only a minor effect on the outwash. The reinforcement lines will come out laterally from the aircraft giving a higher velocity and deeper wall jet somewhere along the side rather than at the front or the back.

IV. Surface Erosion

The ability of the landing surface to withstand the jet's impingement depends on a number of things including the type of surface material, the temperature and pressure of the impinging jet, and how long the surface is subjected to the jet impingement. Most of the investigations of the effect of jet impingement on landing surfaces have been carried out by British Aerospace[4,5,9–11] and by the Naval Air Engineering Center.[12,13]

The different surfaces that have been examined include the following: 1) grassland; 2) concrete; 3) asphalt; 4) aluminum planking (for mobile landing pads), such as AM-2 mat; 5) steel and aluminum plate (carrier decking); and 6) antiskid surfaces applied to steel plate, such as AMSS.

The failure mechanisms and areas of concern for these surfaces are dealt with in turn in the following sections, but the following two plots should be borne in mind through these sections. Figure 6.5 from Ref. 12 shows the surface temperature attained by most of these different surfaces following a 10-s exposure to a heated jet.

Figure 6.6 shows how attained surface temperature varies with time when a jet of 2800°F impinges on the surface. The low thermal conductivity materials reach much higher temperatures more quickly than the high thermal conductivity materials, which helps explain their subsequent behavior.

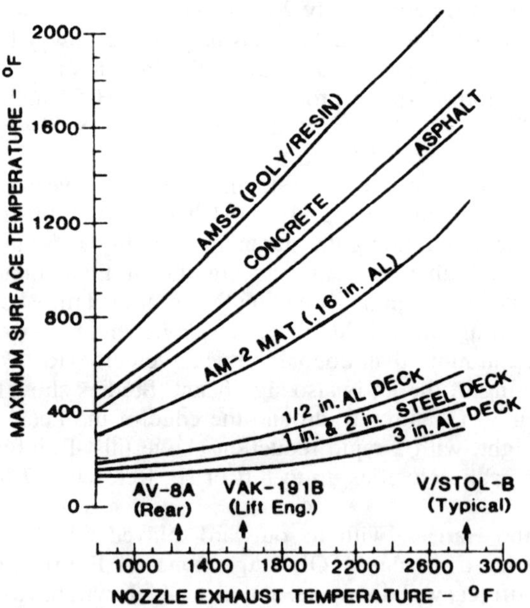

Fig. 6.5 Effect of jet temperature on the maximum surface temperature reached after 10 s.[12]

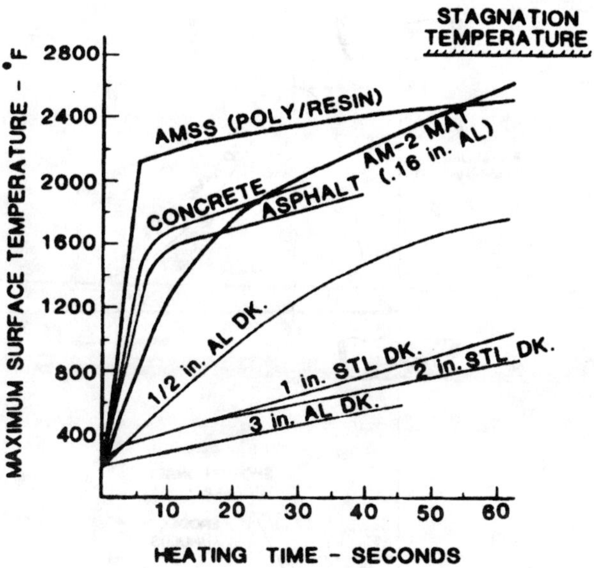

Fig. 6.6 Effect of exposure time on the maximum surface temperature reached on various materials.[12]

A. Grassland

Reference 5 reported early investigations of the static impingement of a single jet during hover. The erosion of grassland began at a point about one diameter from the impingement point, which correlates with the peak value of the dynamic pressure in the radial wall jet (and therefore the peak scrubbing action) as shown in Fig. 6.7. Grassland has not been used for jet V/STOL aircraft because the XV-6A Kestrel demonstrated just how much damage it could do to this surface. Helicopter-like levels of disc loading, with relatively low dynamic pressure in the wall jet, are required to be able to operate from grassland. Reference 1 quotes a peak allowable impingement pressure ratio of between 1.3 and 1.4.

B. Concrete

Concrete is an excellent material for runways for conventional aircraft, but it has limited tolerance for elevated temperatures. Reference 12 reported experiments on the impingement of a jet on a concrete slab 8 in. thick. The study was conducted in support of a concept that potentially used afterburning jets for VTOL operations and could have exhaust temperatures in this regime up to 2800°F. Analysis of the temperatures at various depths in the slab is presented in Fig. 6.8. In each of the plots, the impingement point is at the far left (zero in x), and the point at which the maximum temperature was observed is marked with a star. Reference 12 lists three thresholds for the temperature attained within the concrete: 1) at about 200°F water evaporates leading to

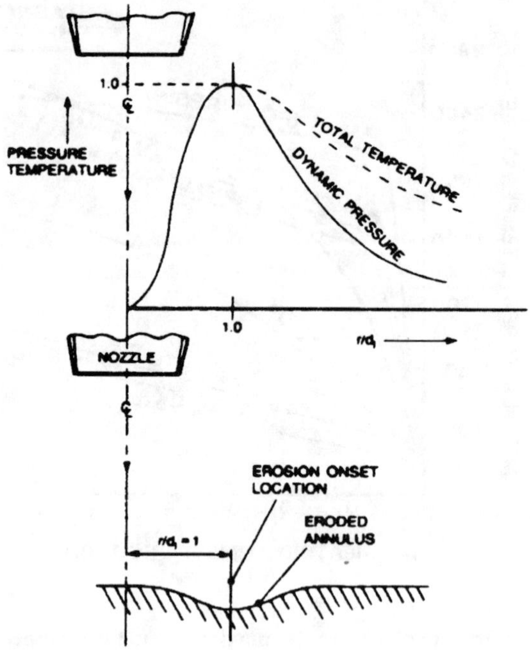

Fig. 6.7 Correlation of erosion onset with peak dynamic pressure—subcritical jets.

hairline cracks and surface flaking; 2) at about 500°F spalling occurs because of the thermal shock of rapid expansion and contraction; and 3) at about 1000°F concrete disintegrates as a result of volume changes in the aggregate ingredients.

The top presentation in Fig. 6.8 shows that after an exposure of 15 s to the full 2800°F exhaust temperature a surface temperature of 1830°F is reached indicating

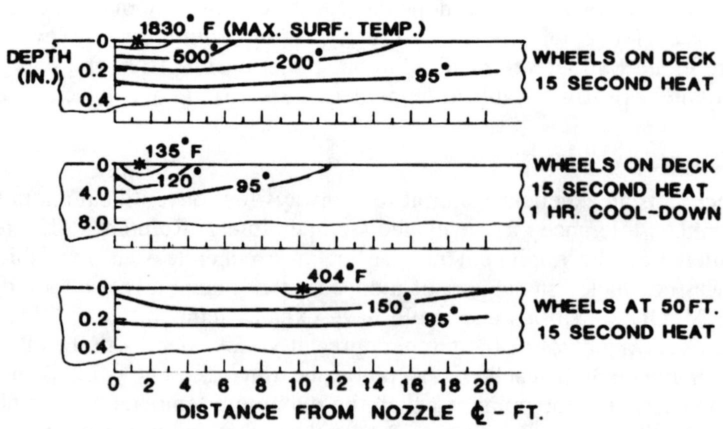

Fig. 6.8 Isotherms in concrete exposed to jet impingement.[12]

that severe surface erosion would occur. Also spalling caused by thermal shock (500°F) would occur over a 15-ft-diam area, and surface flaking (200°F) would extend over a 30-ft-diam area. The center plot shows the temperature profile after a one-hour cooldown. A peak temperature of 135°F is observed. Even this temperature is a source of concern for personnel working under the aircraft.

The lower plot on Fig. 6.8 shows the surface and internal temperatures reached when hovering at an altitude of 50 ft, for example, in preparation for a creeping or rolling vertical landing as already discussed. Even at this altitude 15 s of hover would lead to surface temperatures that would cause surface flaking. The failure mechanism is proportional to the heat flux in the jet, which is a combination of both NPR and temperature.

C. Asphalt

Asphalt roadways are used for STOL operations for Harriers, where the jet residence time is low. The bitumen binder for asphalt softens at relatively low temperatures, and once it does so the particles within the surface are not able to withstand the scrubbing action of the jet. The failure pattern is similar to grassland once the binder is soft. Damage can also be caused by the flowing, or migration, of the binder, even though material might not be lost. Reference 1 estimates a maximum jet temperature for vertical operations of less than 400°F.

D. Aluminum Planking

AM-2 matting is a modular form of planking used to construct temporary landing strips at forward bases. The matting has a corrugated section to give it strength and is as thin as possible for lightness; it being 0.16 in. thick. The high thermal conductivity of the aluminum spreads the thermal load through the matting, but the low volume of aluminum means the impingement point soon approaches the stagnation temperature of the jet. The subsequent weakening of the aluminum leads to the first form of failure, which is a buckling of the planking when mechanically loaded, usually by the aircraft landing on the weakened mat. The mechanical strength of most aluminum alloys degrades quickly beyond a critical temperature in the region of 350°F.

The second, more extreme, form of failure is for the jet to punch a hole through the weakened plank. High-temperature jets will cause the first form of failure, whereas both high-temperature and high-NPR jets are needed for the second form. It is very difficult to put numbers to the temperatures and pressure ratios needed because of the dependence on residence time and the size of the landing pad.

E. Steel Deck

The investigations of Ref. 12 also examined metal plates, both steel and aluminum. Figure 6.9 shows the internal isotherms for three types of metal plates after being subjected to the impingement of the 2800°F jet for 15 s. In all cases the high thermal conductivity of the metals carries heat through to the lower side of the plate so that the air, or other backing material, beneath is heated.

The steel plate is representative of aircraft carrier deck material (top of Fig. 6.9) and reaches a maximum surface temperature of 504°F, which is within acceptable limits. This is for a single 15-s operation. Multiple operations

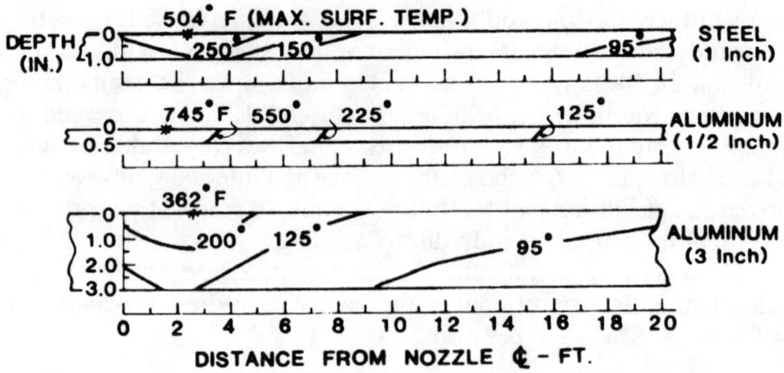

Fig. 6.9 Isotherms in metal plates exposed to 2800°F jet impingement.[12]

would result in higher temperature and could eventually lead to deck failure, probably similar to that inflicted by AV-8A Harriers on the LPH Tripoli (Refs. 12 and 14). In the Tripoli incident 14 Harrier takeoffs from the same spot, in rapid succession, led to buckling of the deck and the formation of an 8-in. depression that required 16 h to cool. This experience and the analysis of Ref. 12 suggest the need to build expansion joints into the landing surface to accommodate the thermal expansion imposed by the jet impingement.

The analysis of the $\frac{1}{2}$-in.-thick aluminum plate (center of Fig. 6.9) shows that the heat penetrates completely through the plate and the isotherms are essentially concentric rings around the impingement point. Increasing the aluminum plate thickness to 3 in. (bottom of Fig. 6.9) significantly reduces the maximum surface temperature, but it needs to be remembered that these results are all for a single 15-s operation.

F. Antiskid Coatings

The antiskid coatings on aircraft carrier decks are typically some form of epoxy resin with a coarse grit-like substance mixed in, which is painted onto the deck. The resin has a low thermal conductivity and so insulates the deck structure to a certain extent. Although this is good for the steel deck, the thermal stresses caused by the hot surface and relatively cool substructure lead to the flaking of the coating and its ablation. The sand-blast effect of eroding antiskid coating is well known to carrier deck crews who operate Harriers.

G. Modeling and Simulation

Computer modeling of ground erosion is heavily dependent on good CFD predictions of heat-transfer coefficients and dynamic pressure and an ability to use these values as boundary conditions to predict the behavior of a nonhomogeneous material. Neither of these tasks is straightforward, and the combination of the two has precluded a successful method of erosion prediction from being developed to date. As both CFD and materials modeling improve, then erosion will undoubtedly become amenable to accurate predictions.

In the mean time, researchers are left with experimental simulation techniques to understand the behavior and limitations of different materials. Although the use of real jet engines would remove question marks about scale effects and "real jet" effects, the cost and complexity of this approach means that only rather few, simple experiments, that is, fixed height, have been conducted using a working engine. The majority of all test work has been carried out at subscale.

British Aerospace (now BAE Systems) developed a subscale rig, Fig. 6.10 specifically for the purpose of furthering the understanding of ground erosion (Ref. 9 and updated in Ref. 11). The rig uses an aircraft engine combustor to produce a jet at temperatures from 500 K (466°F) to 1500 K (2266°F) at pressures up to 10 atm. It is fitted with nozzles up to 12 cm (4.75 in., approx.) diam that can be mounted vertically at heights from 2 to 5 diam above the surface under test. The nozzle can also be mounted horizontally.

The rig has a track under the nozzle assembly, on which a trolley is run using a linear induction motor. This can be accelerated up to a speed of 25 ft/s. Surface samples are 2 ft square and can be mounted on the trolley or fixed beneath the nozzle. For fixed height tests, with the jet vertical, a heat shield is used to cover and protect the surface sample until the desired jet conditions are achieved. The trolley pulls the heat shield away, and the combustion chamber is shut down to terminate the run at the appropriate time. Samples can be passed beneath the vertical jet to simulate aircraft forward motion at a fixed height.

To simulate vertical landing, the nozzle is mounted horizontally, and the sample is mounted vertically on the trolley, which is driven toward the horizontally directed jet at the vertical landing sink rate.

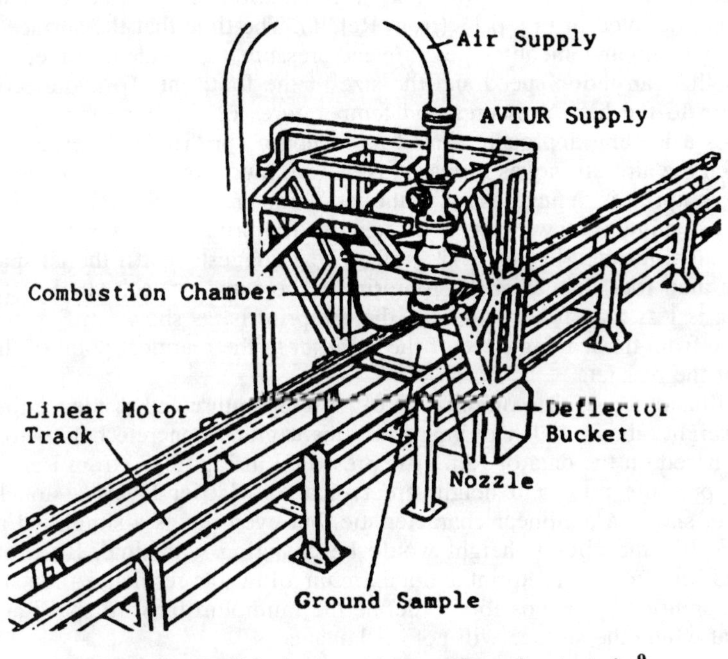

Fig. 6.10 British Aerospace ground erosion test rig.[9]

Assessment of damage, and damage onset, can be tricky for ground erosion. To detect flaking or spalling of concrete, the experimental runs are recorded using high-speed video. Careful viewing of the video is required to determine erosion onset, and the subsequent erosion behavior of the material. Some asphalt erosion does not result in the ejection of material, but in migration of the bitumen binder. This is characterized by digitally scanning the surface of the sample before and after experimental runs and comparing them. For antiskid coatings the friction coefficient of the surface is also assessed before and after experimental runs.

Large-scale experimental investigations of the ability of various materials, ranging from metals to standard concrete to concretes reinforced with metal fibers to withstand high-temperature jet impingement, were also reported in Refs. 12 and 13. These studies were carried out using the exhaust of a J79-8 turbojet engine at military and afterburning power settings, impinging statically, on various potential landing surface materials. In general the experimental results corroborated the studies already discussed, showing that there are no serious scale effects or real engine effects.

H. Options for Reducing Erosion

1. Translation Speed

As already stated, most surfaces of interest for V/STOL operations fail initially because of reaching too high a temperature, which breaks down the cohesion of the material, at which point the dynamic pressure of the jet begins to erode the material. The most effective way to delay this breakdown is to reduce the residence time by translating over the surface.

A definition of residence time and an illustration of the effect of translation speed is presented in Fig. 6.11 (from Ref. 1). The time that the surface is subjected to the impingement temperature and pressure, the residence time, is a function of the translation speed and the size of the footprint. To avoid prolonged exposure to the high pressures and temperatures of the impinging jet, Ref. 1 suggests a landing approach termed a "creeping" or "rolling" vertical landing in which the aircraft moves forward just before touchdown. For a single circular jet the footprint is defined as twice the jet diameter.

For a configuration with jets in tandem, the footprint depends on the spacing of the jets as illustrated in Fig. 6.12. Reference 1 suggests that if the jet spacing is greater than four diameters the footprint is the same as each single jet. If the spacing is less than four diameters, the footprint is, as shown in Fig. 6.12, the distance from the leading edge of the front jet to the rearmost point of the footprint of the rear jet.

An illustration of the combination of jet temperature and residence time, at a fixed height, above which ordinary pavement grade concrete begins to breakdown, based on the data of Ref. 1, is presented in Fig. 6.13 (from Ref. 3). The nozzle pressure ratio and height are typical of Harrier levels at touchdown. The plot shows a nonlinear characteristic, but even this is a simplified picture, as in reality the aircraft height would be changing with time. By combining the data for effective footprint and maximum allowable residence time at touchdown conditions, it is possible to define the minimum forward speed at touchdown at which the surface will not be damaged.

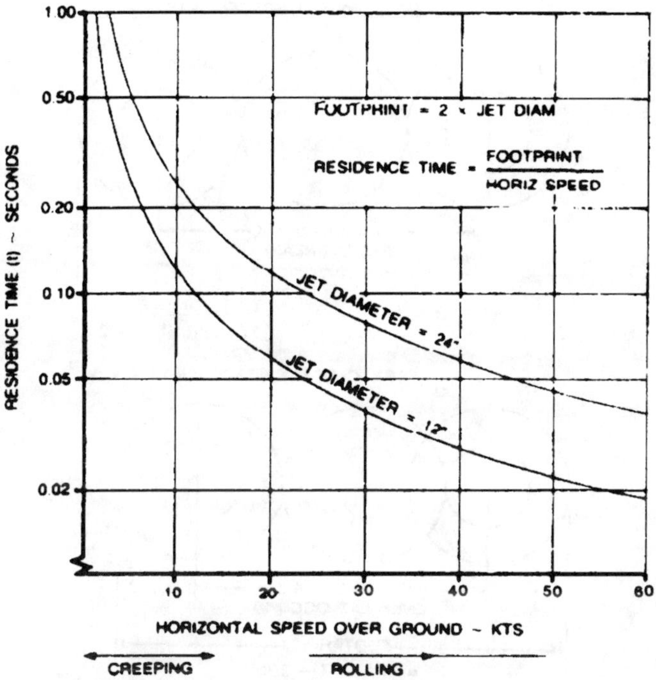

Fig. 6.11 Effect of translation speed on residence time.

2. Rapid Mixing Nozzles

A means of reducing both the dynamic pressure and temperature of the impinging jet, and thus alleviating erosion potential, would be to increase the mixing rate, or decay of the jets. The mixing of the edges of the jet stream with the surrounding air causes the jet velocity to decay with distance from the nozzle, as shown in Fig. 2.1. An investigation into the possibility of developing nozzles that would have a more rapid rate of decay is reported in Ref. 15. The approach was to break the nozzle into various arrangements of smaller nozzles. Six of the 12 nozzle configurations tested in that investigation are shown in Fig. 6.14.

Figure 6.15 shows that significant reductions in both temperature and dynamic pressure can be achieved. The dynamic pressure reduction achieved by a circular nozzle at a distance of 10 diam from the nozzle can be achieved by some multiple nozzles at only two diameters. Temperature reductions are a little smaller.

These reductions in dynamic pressure and temperature, however, come at a price. Figure 6.16 shows the effective velocity coefficient C_v for the nozzles that produce large reductions in impingement dynamic pressure have thrust losses of 4 to 8% in the nozzle itself. $C_v = V_e/V_i$, where V_e is the effective velocity, based on the measured thrust Th and mass flow m, $V_e = Th/m$, and V_i is the ideal velocity associated with the nozzle pressure ratio.

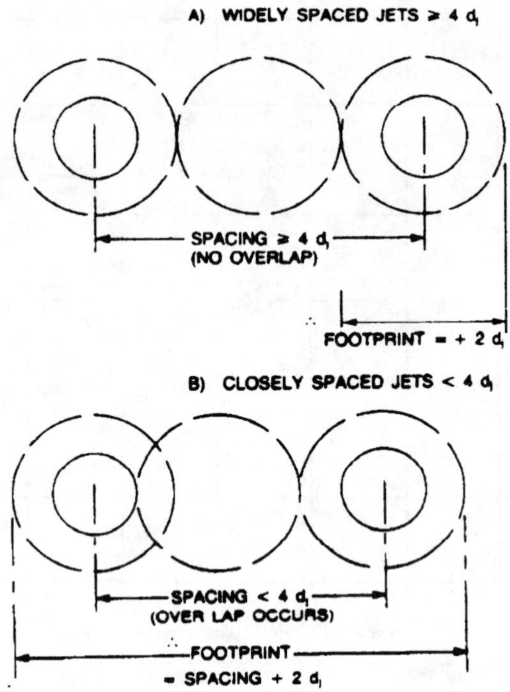

Fig. 6.12 Definition of effective footprint size.[3]

In addition to the basic nozzle losses already discussed, the rapid entrainment that causes the rapid decay shown in Fig. 6.15 also induces increased suction pressures on the area surrounding the jet. These suction pressures cause an additional lift loss as discussed in Sec. 2.1. As shown in Fig. 2.3, this lift loss is proportional to the maximum rate of change of dynamic pressure with distance from the nozzle, is inversely proportional to the distance from the nozzle at which

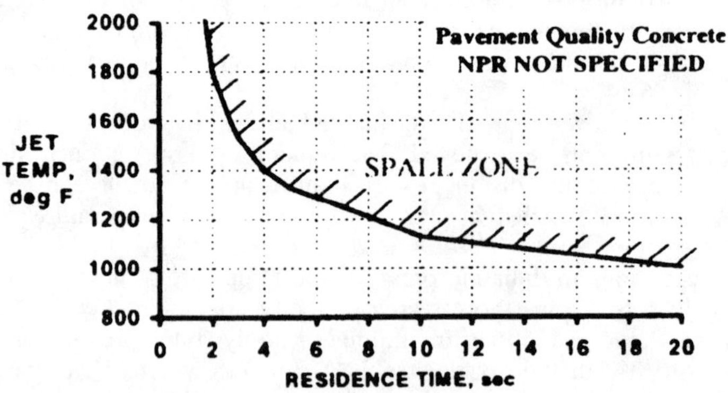

Fig. 6.13 Effect of temperature on the residence time before breakdown of concrete.[3]

CIRCULAR NOZZLE 1.1

NOZZLE 2.2

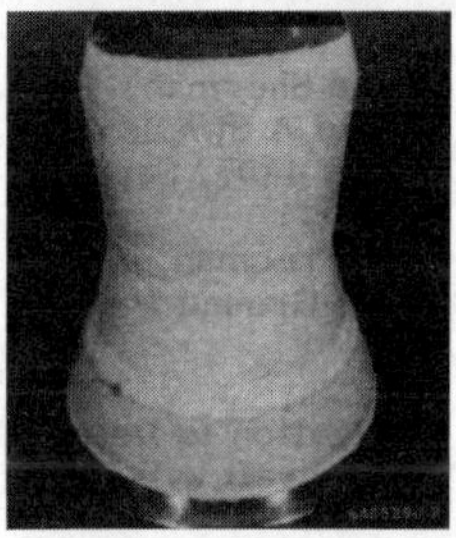

DELTA NOZZLE 1.2

NOZZLE 2.5

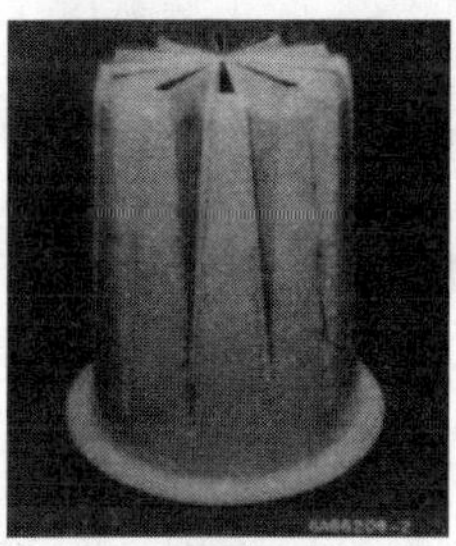

TWELVE SEGMENT NOZZLE 1.3

NOZZLE 2.8

Fig. 6.14 Some of the rapid mixing nozzles investigated.[15]

this maximum occurs, and is proportional to the square root of the ratio of planform area affected to the nozzle area. For the nozzles of Fig. 6.15 that produce the most rapid decay, this lift loss could be as much as 8 to 10% of the thrust, assuming the planform-to-jet-area ratio is of the order of 100.

V. Spray

Shipboard-based V/STOL aircraft usually come to a hover, or near hover, alongside their home ship prior to landing and even helicopters (Fig. 6.17) can generate a considerable amount of spray in this maneuver.

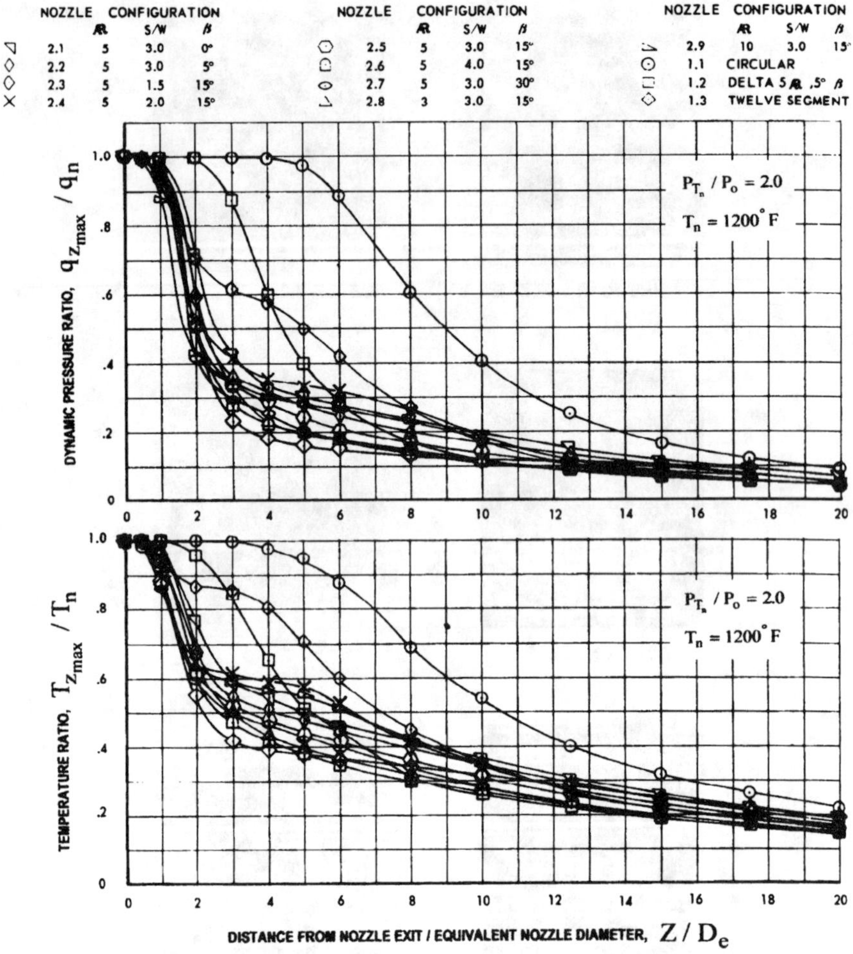

Fig. 6.15 Temperature and dynamic pressure reductions achieved.[15]

Spray is produced by wind blowing over the surface of the water creating a succession of outward flowing wavelets, which in this case comes from the wall jet. The survey of the literature presented in Ref. 16 shows (Fig. 6.18) that spray is not produced until the energy of the outward-flowing wall jet exceeds a dynamic pressure q of about 2 psf, which is low even relative to helicopter disc loading.

As the disc loading and total pressure of the flow impinging on the surface increases, another factor comes into play. At high pressures the impinging stream depresses the surface of the water (right sketch of Fig. 6.18), and a pronounced cavity is formed on the water surface. The slope of the cavity walls gives the spray a direct upward component of velocity. At the lower pressures, but above 2.0 psf, the impingement region is large relative to the depression of

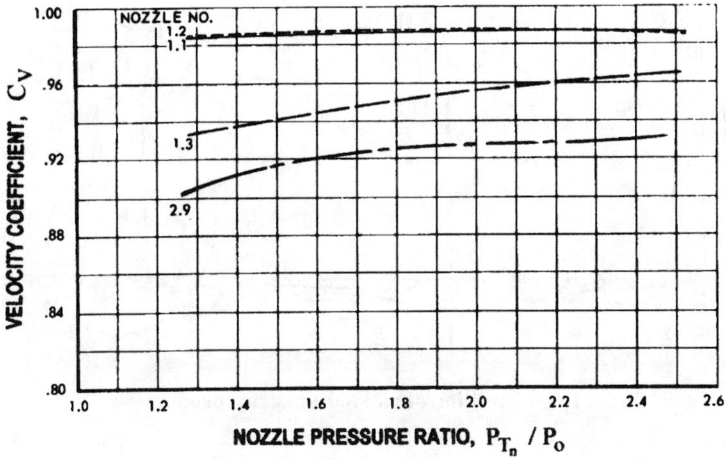

Fig. 6.16 Comparison of the performance of selected nozzles.[15]

the water surface, and the outward flow remains relatively flat. The spray does not have a direct upward component, but is acted upon by the surrounding flowfield.

During normal aircraft operations, limiting the hover to heights equal to or greater than the spray height is desirable to minimize aircraft corrosion and pilot visibility problems. The analysis of Ref. 15 found that the height to which spray was projected by the impingement of streams from jet and fan-powered aircraft were only weakly dependent on the nozzle pressure ratio and that the minimum height at which these aircraft can hover and stay above the spray is approximated by $H_{min} = 0.6\ \sqrt{W}$, where H_{min} is the minimum hover height and W is the weight of the aircraft. The F35B JSF is in the 40,000 lb category, so that this would give an H_{min} of 120 ft above the water.

Fig. 6.17 Spray produced by a large helicopter.[16]

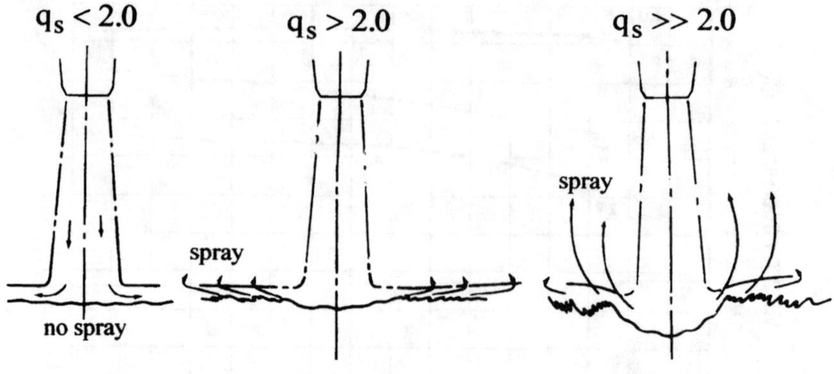

Fig. 6.18 Conditions involved in spray production.[16]

VI. Acoustics

The noise generated by the jets, and the effect of ground proximity in increasing that noise, can be of concern for several reasons. As pointed out in Ref. 1, high noise levels can have an adverse effect on 1) the aircraft structure and external stores, 2) the well being and efficiency of the ground crew, and 3) the crews ability to communicate. Out-of-ground effect, a single free jet (upper left of Fig. 6.19) generates somewhat higher noise levels in the aft quadrant (or arc) than in the forward quadrant.

As the ground is approached, the noise in the aft arc is gradually cut off, but the reflected, or mirror image noise as depicted in Fig. 6.20, is added, causing the peak noise to tend to shift from the aft quadrant toward the forward arc (Fig. 6.19).

The effect of height above the ground on the noise level increment caused by jet impingement, as measured on the aft fuselage of a small-scale twin jet model, is shown in Fig. 6.21. As indicated in Ref. 1, the agreement between the full-scale Dornier DO 31 data and these small-scale data are probably fortuitous, but both sets of data indicate that the effects of the ground are significant and should be considered in the aircraft structural design.

A semi-empirical method for estimating the effects of ground proximity on the noise field of a hovering jet VTOL aircraft is presented in Ref. 17. In free-flight, out-of-ground effect, the primary source of noise from a subsonic jet is the broadband noise generated by the mixing within the core flow region and in the shear layer between the jet and the ambient air.

The internally generated noise from the engine combustion and turbulence, called core noise, can also add a small increment to the noise level. Core noise is much lower than the mixing noise in the static thrust, or hover, condition and only becomes significant at forward speeds, where the lower differential between the jet speed and the surrounding air reduces the mixing and associated mixing noise.

The methods developed for estimating the out-of-ground effect broadband mixing noise (Refs. 17 and 18) agree well with the data from a J85 turbojet (Figs. 6.22 and 6.23) in terms of both frequency spectra and directionality.

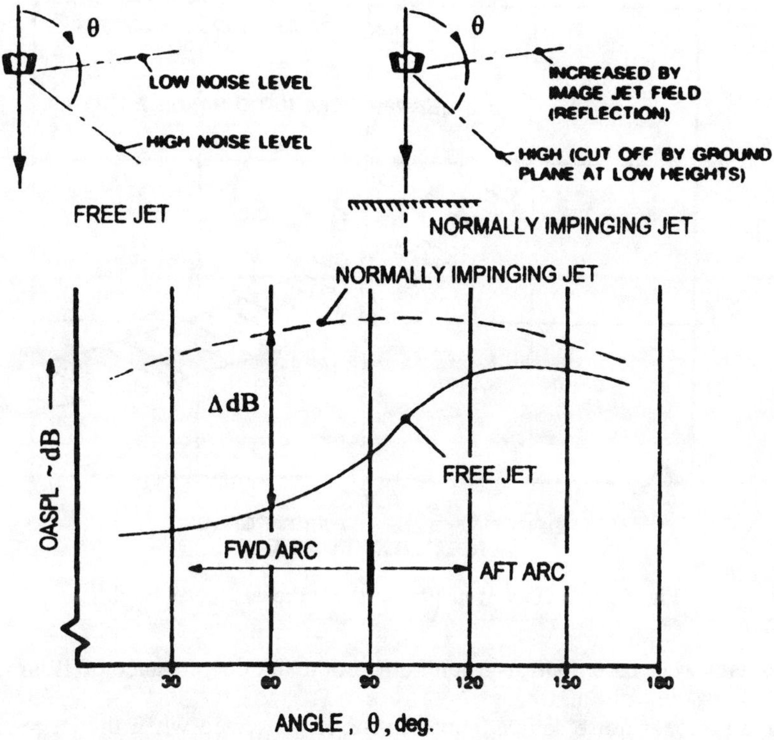

Fig. 6.19 Effect of impingement on the directionality of jet noise.[1]

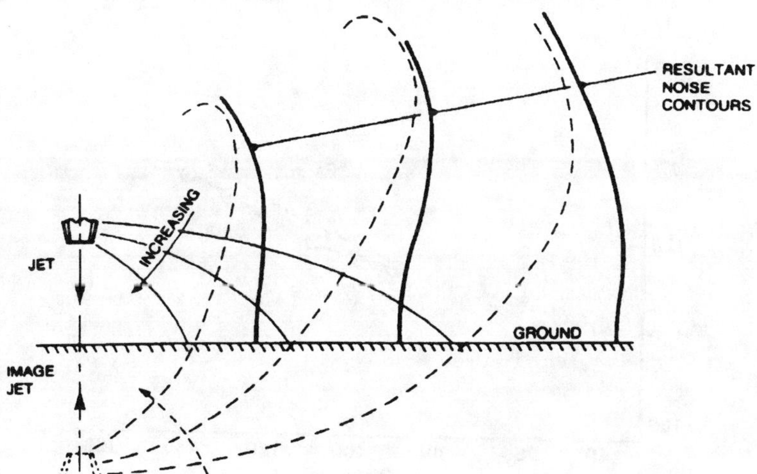

Fig. 6.20 Image reflection theory of jet impingement noise.[1]

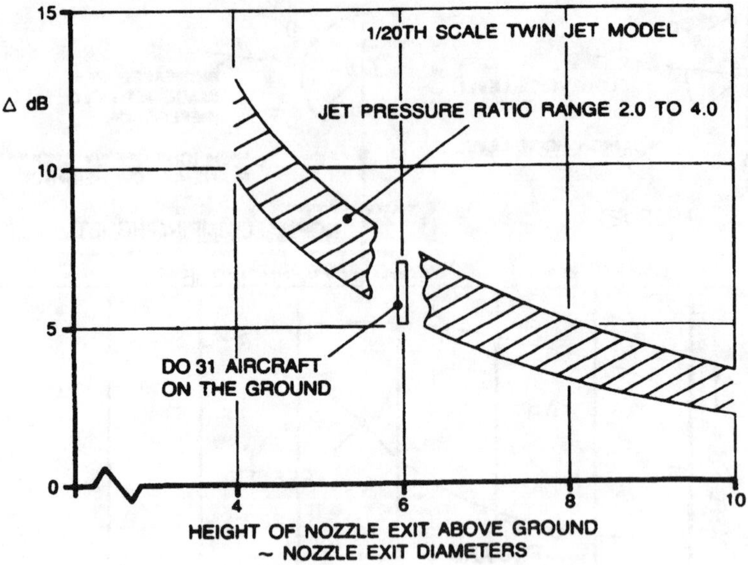

Fig. 6.21 Effect of height on the increased noise level.[1]

These data were taken out-of-ground effect at a sideline distance of 12 m and a nozzle pressure ratio of 1.83.

An additional noise source, shock noise, is generated when the jet pressure ratio is supercritical. The phenomenon is complex because of the many possible interactions between the jet flow and the shock waves in the supersonic plume and has been the subject of many investigations (Refs. 19–27 for example). Figure 6.24 shows a comparison of the predicted components of shock and

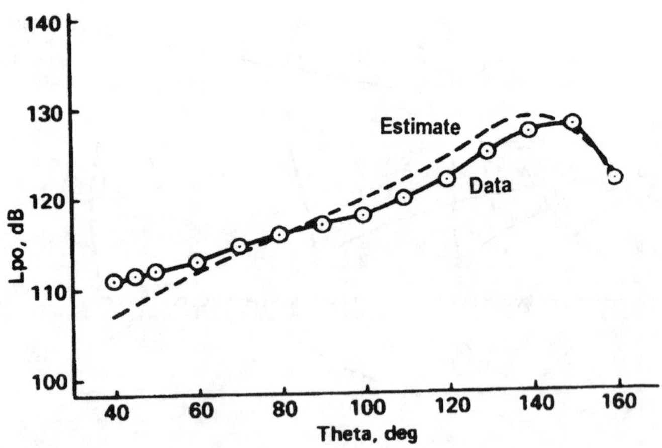

Fig. 6.22 Comparison of the predicted and measured noise level of the J85, out-of-ground effect.[17]

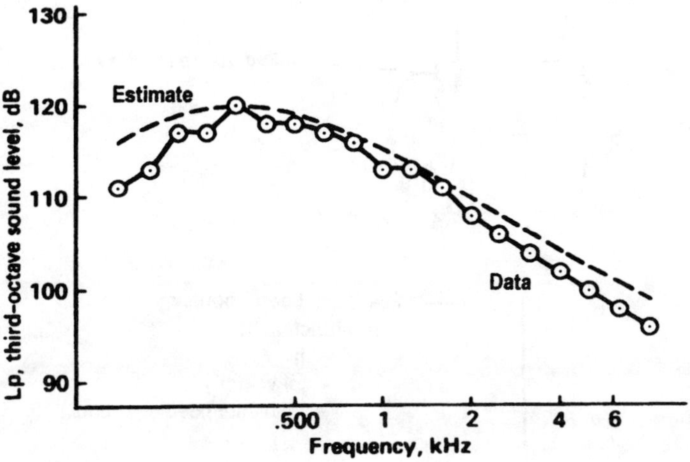

Fig. 6.23 Comparison of predicted and measured third-octave noise, at theta = 140 deg, for the J85.

mixing noise compared with the measured spectrum for a 1.3-in.-diam hot jet at a pressure ratio of 3.0. These results indicate that the shock noise predominates in the forward quadrant but is dominated by the mixing noise in the aft quadrant.

During hover, proximity to the ground increases the noise felt by the airframe and in the far field over that experienced out-of-ground effect. Three factors are at work:

1) There is an acoustic reflection (Fig. 6.20) by the ground.

2) The jet noise sources are altered by aeroacoustic interactions with the ground (Fig. 6.25).

3) New noise sources are generated in the ground wall jets flowing outward from the jet impingement points.

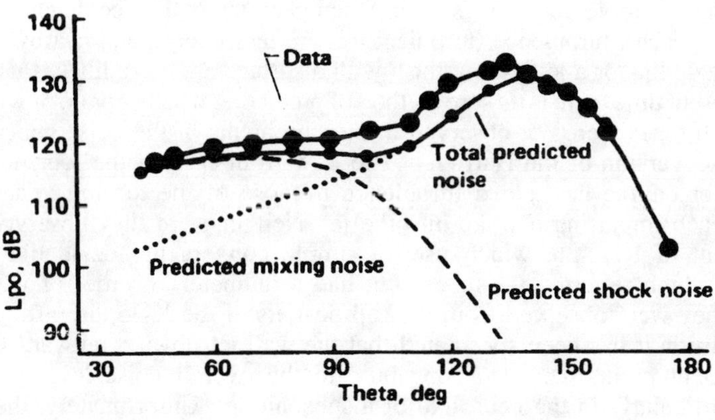

Fig. 6.24 Comparison of the sum of the predicted shock and mixing noise with data for a supercritical hot jet.[17]

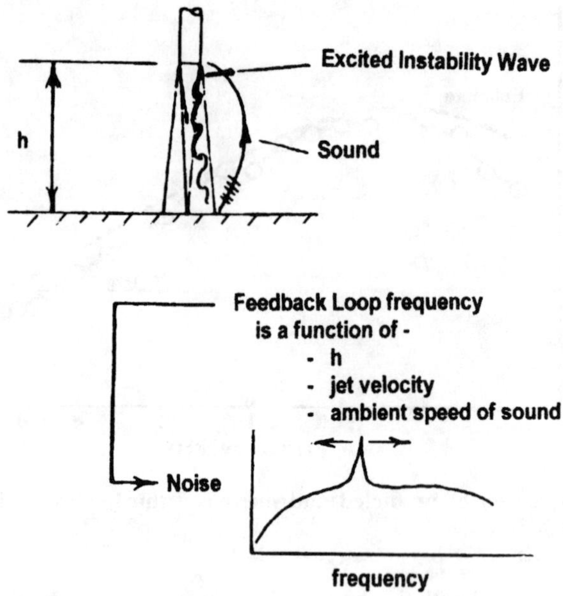

Fig. 6.25 Impingement tone generation mechanism.[20]

Intense tones, commonly referred to as screech, have been observed in some small-scale investigations of the noise produced by a jet impinging on the ground (Refs. 15–17, 20, and 27, for example). As stated in Ref. 27, "the aeroacoustic mechanism is driven by ring vortices in the jet which strike the ground, radiate sound back to the nozzle to excite further vortex rings, and thereby create a resonant, coherent flow structure and acoustic tones. However the aeroacoustic resonance, as is true of all aeroacoustic resonances, is sensitive to the Reynolds number of the flow. Turbulence can disrupt the resonance mechanism."

Screech is a very sensitive phenomenon, dependent on factors such as scale (small scale is more likely to screech), temperature of jet (cold jets are more likely to screech), turbulence (turbulent jets are less likely), and nearby structure (thick nozzle lips or a jet issuing flush with a surface are more likely to screech). At the present time there is no sure method of predicting whether or not it will occur.

Screech tones were not observed in measurements of the noise generated by the AV-8C version of the Harrier.[25] The flow from each of the four nozzles of the Harrier engine has added turbulence because of the turning vanes in the nozzles each imparting a wake into the jet. Neither were they observed in the experiment of Ref. 26, which used a simple convergent nozzle like that of Ref. 20 (which reported the tones) but had a diameter an order of magnitude larger. They were observed from the roll postjets of the X-35 aircraft.

In addition it has been speculated that the wall jet flowing outward from the impingement points can also contribute to the overall noise level in ground effect, particularly in the near field of a subsonic jet. Unfortunately, there have been few investigations directed at determining the contribution of these wall jets to the overall noise level and no method for estimating this contribution is available.

References 17 and 25 report on a flight investigation of an AV-8C Harrier aircraft[25] undertaken to develop a database for the acoustic environment during vertical takeoff, vertical landing, and hover out-of-ground effect. As shown in Fig. 1.2, the engine is fitted with four nozzles. The fan bypass flow is exhausted through the two front nozzles, and the turbine hot gas is exhausted through the two rear nozzles. Two sets of turning vanes are used in each nozzle to achieve good turning of the flow in this closely coupled configuration. The nozzles are basically rectangular in cross section, but, on the engine used in the AV-8C investigation of Ref. 17, both sets of nozzles are scarfed, that is, are cut parallel to the trailing edge of the turning vanes, not perpendicular to the axis of the jet. This can be compared to the engine shown in Fig. 1.3, where the rear jets only are scarfed.

The investigation started with out-of-ground-effect measurements. Figure 6.26 shows the directivity pattern, in the plane 90 deg to the aircraft centerline, for selected third-octave frequencies. This pattern is different from that expected for a single isolated freejet as estimated from Ref. 18. Reference 17 speculates that the Harrier pattern is closer to omnidirectional than a single isolated jet would be because of the reflections from the wing and fuselage.

The ground amplification of the noise environment of the AV-8C Harrier is presented in Fig. 6.27 for takeoff and landing power settings for a sideline distance of about 42 ft. Takeoff and climb was made at full power, whereas

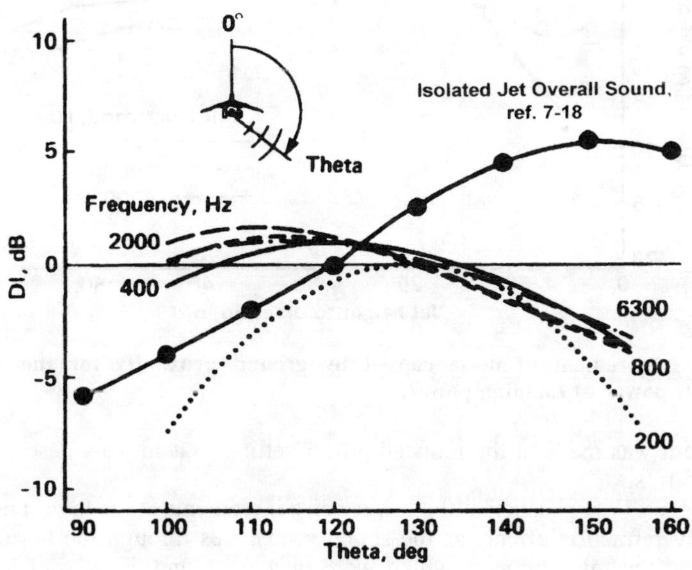

Fig. 6.26 Directivity pattern of the AV-8C Harrier jet noise, hovering out-of-ground effect.[17]

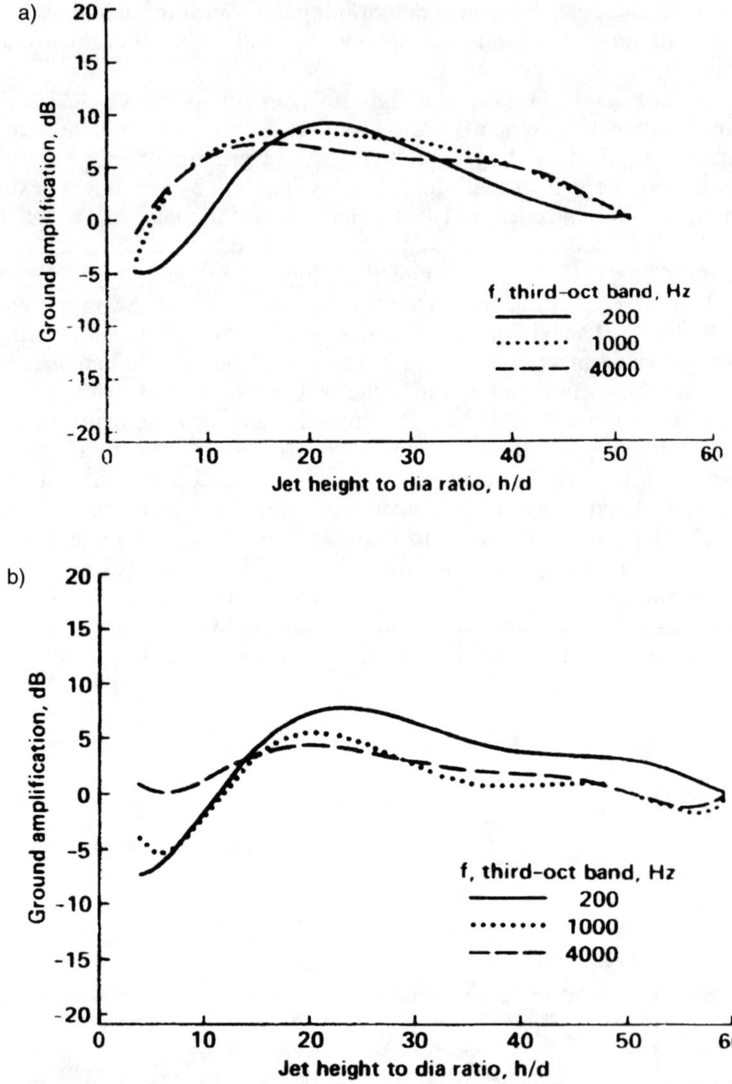

Fig. 6.27 Increment of noise caused by ground proximity for the AV-8C.[17] a) Takeoff power b) Landing power.

the descent was made at the reduced power setting to achieve a descent rate of about 2.5 ft/s.

Reference 17 suggests that the lower noise increments at the lowest heights is caused by refraction effects as the sound propagates through the hot turbulent wall jet to the microphones, which were on the ground. That wall jet would cause sound waves to refract upward. The refraction would be strong for sound propagating parallel to the ground (low-altitude jet) and weak for sound propagating normal to the ground (high-altitude jet). Thus the sound heard at some

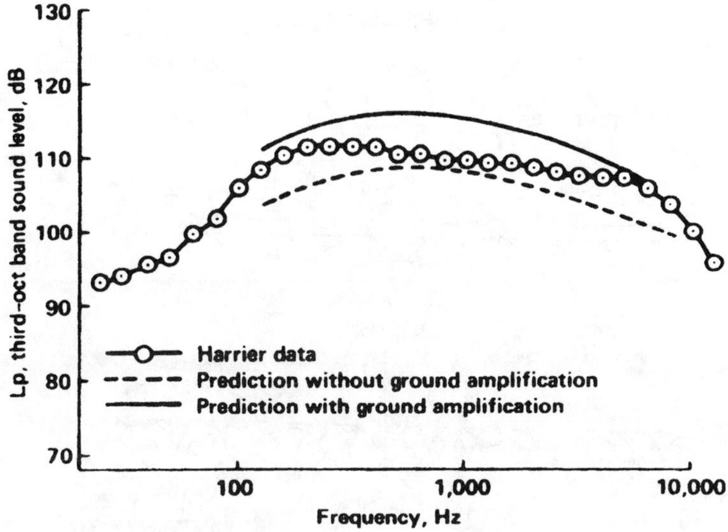

Fig. 6.28 Comparison of predicted noise level with measured data at an angle of theta = 21 deg for AV-8C at $h/d = 28$.

distance above the ground could actually be maximum with the jet near the ground.

A comparison of the noise measured on microphone #4, about 100 ft to the side with the aircraft, at an altitude of about 40 ft, with that predicted by the semi-empirical expressions developed in Ref. 17 is presented in Fig. 6.28. The agreement is reasonably good. Reference 17 suggests that improvements could be made in the prediction if configuration effects such as nozzle shape and proximity to the airframe or multiple jet effects could be better accounted for in the method. A computer code for the method is presented in Ref. 17.

VII. Ground Surface Modifications

There have been several investigations of ways to modify the landing surface to reduce adverse ground effects. These surfaces could be useful either for shipboard or for remote site applications. The most extensive of these studies are reported in Refs. 28–31. These studies refer to their concept as ground environment mat (GEM). The operating concept is portrayed in Fig. 6.29. With the GEM (lower sketch in Fig. 6.29) a porous surface is supported above the fixed ground at a height such that the radially flowing wall jet can penetrate the porous surface and flow outward between this and the solid ground. The aerodynamic performance of the GEM is controlled by three geometric parameters: 1) porosity of the top surface; 2) vertical space between the top surface and the solid base; and 3) internal supports, which become drag elements to the radial wall jet. Other factors include the GEM size, perimeter treatments, etc.

Several GEM designs have been studied. Two recent versions[31] are sketched in Fig. 6.30. The GEM P2 design (left side of Fig. 6.30) has a top porous surface

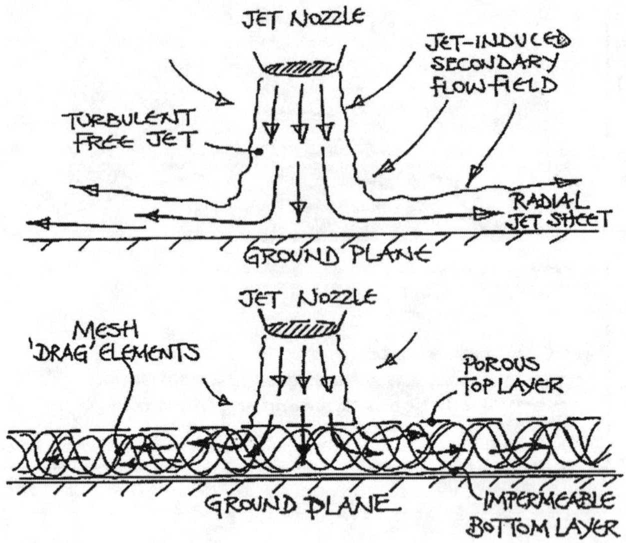

Fig. 6.29 Fluid flow of jet impingement on a solid ground surface and on a modified ground surface (GEM).[28]

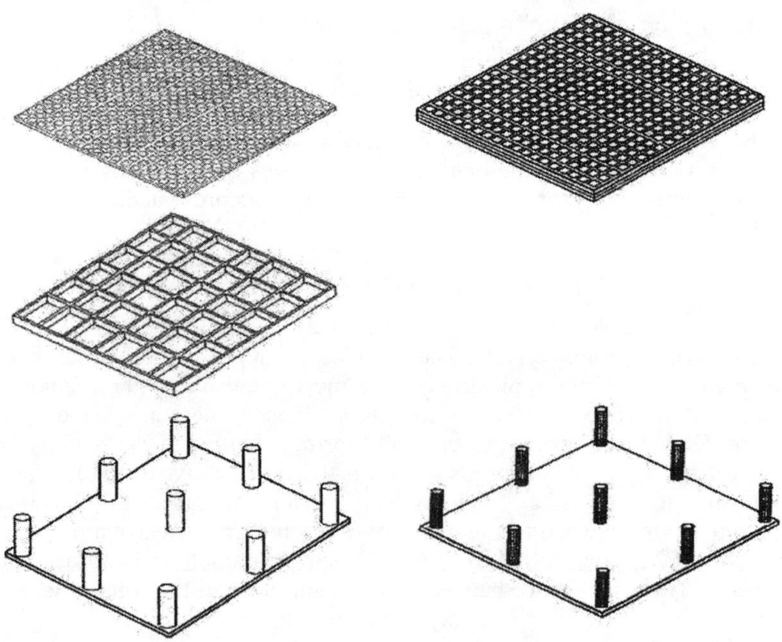

Fig. 6.30 Key components of two GEM designs.[31]

of about 50% open area, a lattice structure underneath it, and supports. The P3 design (right side of Fig. 6.30) has a single thick top porous surface of about 60% open area and supports. Although the P3 is a simpler design for fabrication and maintenance, it is heavier than the P2 design. Tests have not been done to assess the robustness of these designs in an operational environment.

In an earlier investigation[28] several alternative designs of GEM were evaluated. There was a light GEM (GEM 1: a lightweight, transportable mat for temporary use), a heavy GEM (GEM 2: a more permanent installation for either land- or ship-based use), and a collapsible GEM (GEM 3: another mat suitable for either land-or ship-based use). The effect of the GEM design on the ground-effect lift loss throughout the height range tested is presented in Fig. 6.31. There are two data curves presented where the GEM is not used. The curve labeled "solid floor" was measured using the basic three-poster ASTOVL (Advanced Short Take Off Vertical Landing) configuration. The curve labeled "with CAD" was measured using the ASTOVL configuration with a lift improvement device or fence deployed on the rear fuselage lower surface. There is a positive jet-induced lift over the range of ground heights. However, in the VTOL mode the aircraft still incurs the out-of-ground-effect lift loss of a few percent of the thrust independent of the GEM design.

All of the GEM mats tested showed substantial sound-pressure reduction. The average reduction ranges from 3 to 10 dB depending on the GEM design and jet conditions, or between 6 and 9 dB for the best performing GEM over the range of jet conditions. Figure 6.32 shows the effect of GEM in attenuating

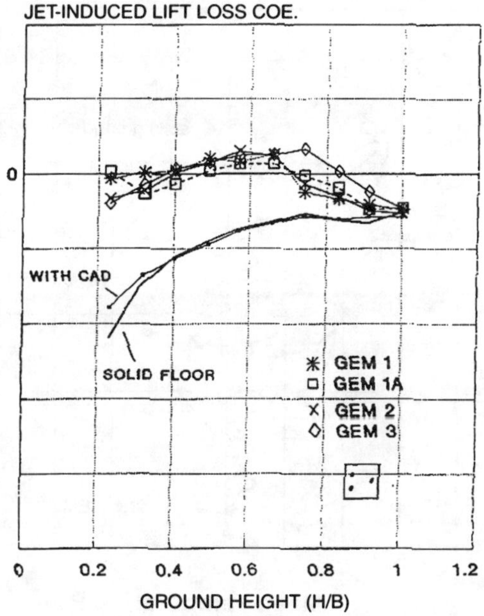

Fig. 6.31 Effect of GEM surface on lift loss.

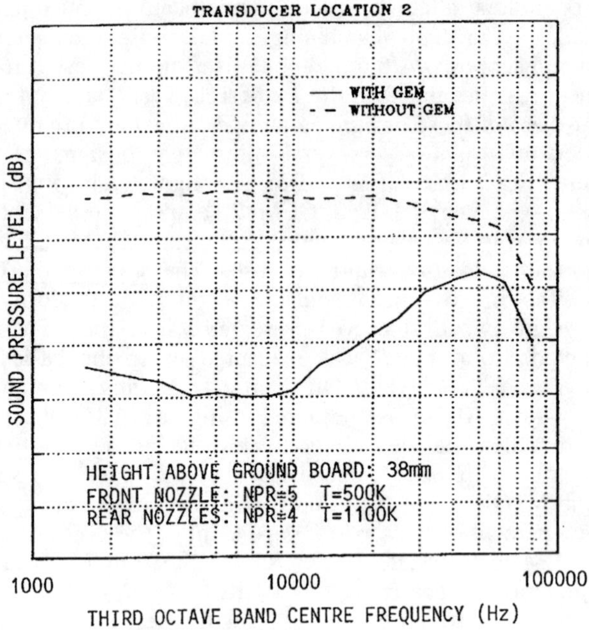

Fig. 6.32 Effect of GEM surface on noise spectra.

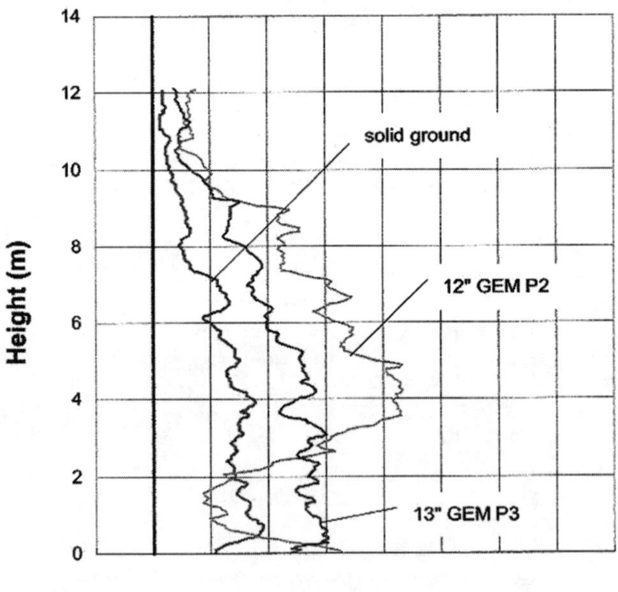

Fig. 6.33 Effect of GEM surface on inlet temperature rise.

the sound-pressure level on the fuselage lower surface. Note that the typical resonance frequencies of an aircraft panel are near the midrange of these measurements. For both acoustic and temperature attenuations, GEM was found to perform better at the higher nozzle-pressure-ratio conditions.

Measurements of inlet temperature rise (HGI) were made to evaluate the effect of the GEM. The flow suppression properties of the GEM caused an accumulation of slow-moving hot exhaust above the GEM surface, which resulted in an increase in the HGI level at most of the ground heights tested and can also increase the surface temperature of the aircraft undersurface. An example of HGI data[31] is presented in Fig. 6.33, where the effects of two GEM designs (P2 and P3) are presented for a Harrier II model with a 12-kn headwind. The HGI problem is aircraft configuration dependent and remains a potential risk to be addressed in future GEM studies before any operational application.

Nomenclature

AR = aspect ratio, D^2/area or length/width
C_v = velocity coefficient
D, D_e = diameter of circular jet of equal area, ft
DI = directivity, dB
d, d_j = jet exit diameter, ft
F = temperature, °F
$H_{\min}$ = minimum hover height to stay above spray, ft
h = height above surface, ft
L = length of an element of rectangular exit nozzle, ft
Lp = third-octave sound pressure level, dB
Lpo = overall sound pressure level, dB
M = mach number of jet flow at exit
P_r, NPR = nozzle pressure ratio
P_{T_n} = jet total pressure at the nozzle, lb/ft^2
P_0 = ambient pressure, lb/ft^2
q_n = jet dynamic pressure at the nozzle exit, lb/ft^2
q_s = dynamic pressure in the wall jet flowing along the surface, lb/ft^2
$q_{z_{\max}}$ = maximum dynamic pressure at distance Z from nozzle exit, lb/ft^2
r = radial distance from center of jet impingement, ft
S = distance between centerlines of ground plane, ft
T_n = jet temperature at nozzle exit, °F
$T_{z_{\max}}$ = maximum temperature at distance z from exit, °F
t = residence time, s
W = aircraft weight, lb; or width of an element of rectangular exit nozzle, ft
z = distance from the nozzle exit, ft
β = nozzle wall divergence angle, referred to the longitudinal axis of the nozzle, deg
Θ = angle measured from jet axis, 180 deg is in the jet direction, deg

References

[1]Knott, P. G., "The Ground Environment Created by High Specific Thrust Vertical Landing Aircraft," SAE P-203, *Proceedings of the International Powered Lift Conference*, 1987, pp. 75–86.

[2]Gummer and Hunt, "The Impingement of a Uniform Axisymmetric Supersonic Jet on a Perpendicular Flat Plate," *Aeronautical Quarterly of the Royal Aeronautical Society*, vol. 22, 1975, pp. 403–420.

[3]McCarthy, K. M., "An Overview of JSF STOVL Performance and Basing Interface Analysis," SAE P-309, *International Powered Lift Conference Proceedings*, 1997, pp. 119–127.

[4]Ghee, T., Gonzalez, H. A., Lake, R. E., and Naumowicz, T., "Velocity and Temperature Measurements of an AV-8B Harrier Wall Jet During Hover," *Royal Aeronautical Society International Powered Lift Conference*, 1998, pp. 14.1–14.14.

[5]Dent, J. M., "Overcoming Ground Erosion Effects on the Operation of Jet Lift Aircraft," AGARD, Rep. 516, Oct. 1965.

[6]Rubel, A., "Oblique Impingement of a Round Jet on a Plane Surface," *AIAA Journal*, Vol. 20, No. 12, 1982; also AIAA Paper 82-4292, 1982.

[7]Miller, P., and Wilson, M., "Wall Jets Created by Single and Twin High Pressure Jet Impingement," *Aeronautical Journal of the RAeS*, Vol. 97, No. 963, 1993, pp. 87–100.

[8]Preston, J. R., "VTOL Downwash/Outwash Operational Effects Model," American Helicopter Society, May 1994.

[9]Wake, A. J., Hill, C. J., and Angel, R. G. A., "Ground Surface Erosion—British Aerospace Test Facility and Experimental Studies," *Royal Aeronautical Society International Powered Lift Conference*, 1990, pp. III.11.1–III.11.11.

[10]Illston, L. V., Angel, R. G. A., Patience, D. E., and Burns, R. E., "Assessment Methods for near Field Noise and Ground Erosion," AIAA paper 93-4870-CP; also *AIAA 1993 International Powered Lift Conference*, 1993, pp. 208–219.

[11]Efford, M. P., Liston-Smith, P., and Tognarelli, R., "The Development and Implementation of a New Surface Erosion Measurement Technique" Society of Automotive Engineers, Paper 2005-01-3173, Oct. 2005.

[12]Fluk, H., "Landing Surface Characteristics Unique to V/STOL Aircraft," SAE P-203, *Proceedings of the International Powered Lift Conference*, 1987, pp. 87–99.

[13]Fluk, H., "Runway Pavement Temperatures: Vertical/Short Takeoff and Landing Operations," Naval Air Engineering Center, NAEC-MISC-903-29, Jan. 1981.

[14]Fozard, J. W., *The British Aerospace Harrier-Case Study in Aircraft Design*, AIAA Professional Study Series, AIAA, New York, July 1978.

[15]Higgins, C. C., Kelly, D. P., and Wainwright, T. W., "Exhaust Jet Wake and Thrust Characteristics of Several Nozzles Designed for VTOL Downwash Suppression," NASA CR-373, Jan. 1966.

[16]Kuhn, R. E., "Height of Spray Produced by Vertical Takeoff and Landing (VTOL) Aircraft," DTNSRDC/ASED-79/04, April 1979.

[17]Soderman, P., "The Prediction of STOVL Noise—Current Semi-Empirical Methods and Comparisons with Jet Noise Data," *Royal Aeronautical Society International Powered Lift Conference*, 1990, pp. III.12.1–III.12.30; also NASA TM 102833, April 1990.

[18]Society of Automotive Engineers Committee A-21, "Gas Turbine Jet Exhaust Noise Prediction," ARP 876B, June 1981.

[19]Krothapalli, A., "Discrete Tones Generated by an Impinging Underexpanded Rectangular Jet," *AIAA Journal*, Vol. 23, No. 12, 1985.

[20]Ahuja, K. K., and Spencer, D. A., "Aeroacoustics of Advanced STOVL Aircraft Plumes," SAE P-203, *Proceedings of the International Powered Lift Conference*, 1987, pp. 531–541.

[21]Kibens, V., Saripalli, K. R., Wlezien, R. W., and Kegelman, J. T., "Unsteady Features of Jets in Lift and Cruise Modes for VTOL Aircraft," SAE P-203, *Proceedings of the International Powered Lift Conference*, 1987, pp. 543–552.

[22]Groen, D. S., "STOVL Acoustic Fatigue Technologies," SAE P-203, *Proceedings of the International Powered Lift Conference*, 1987, pp. 553–562.

[23]Ladd, J. A., Bower, W. W. and Wishart, D. P., "Prediction of Acoustic Loads for STOVL Aircraft," SAE P-306, *International Powered Lift Conference Proceedings*, 1997, pp. 409–423.

[24]Krothapalli, A., Soderman, P. T., Allen, C. S., Hayes, J. A., and Jaeger, S. M., "Effects of Forward Flight on the far-Field Noise of a Heated Supersonic Jet," AIAA paper 96-1720, May 1996; also *AIAA Journal*, Vol. 35, No. 6, 1997, pp. 952–957.

[25]Soderman, P. T., and Foster, J. D., "Noise of the Harrier in Vertical Landing and Takeoff," *1987 Ground Vortex Workshop*, NASA CP 10008, April 1987, pp. 167–190.

[26]Preisser, J. S., and Block, P. J. W., "An Experimental Study of the Aeroacoustics of a Subsonic Jet Impinging Normal to a Large Rigid Surface," AIAA Paper 76–520, July 1976.

[27]Soderman, P. T., and Allen, C. S., "On the Scaling of Small-Scale Jet Noise to Large Scale," NASA TM-103921; DGLR/AIAA Paper 92-02109, May 1992.

[28]Ing, D. N., and Harris, A. E., "Ground Environment Mat (GEM) on ASTOVL Ground-Effect Performance," AIAA paper-93-4891, Dec. 1993.

[29]Ing, D. N., Knott, P., Clark, R., and Appleyard, G., "Ground Environment Mat (GEM): An Assessment of its Performance and Operational Effectiveness For VSTOL Aircraft," *RAeS IPLC Proceedings*, London, 1998.

[30]Ing, D. N., Clark, R., Knott, P., and Appleyard, G., "Ground Environment Mat (GEM): A Recent Assessment of its Design, Performance and Operational Effectiveness For VSTOL Aircraft," IPLC, Oct. 2000.

[31]Ing, D. N., R., Knott, P., and Appleyard, G., "Ground Environment Mat (GEM): Recent Studies of its Design and Performance and For STOVL Aircraft," AIAA paper 2002-5961, Nov. 2002.

[32]Curtis, P., "A Review of the Status of Ground Effect/Environment Technologies," AIAA paper 2002-5985, Nov. 2002.

Chapter 7

Application of Computational Fluid Dynamics

I. Introduction

CONVENTIONAL AERONAUTICS has used a combination of experiment, theoretical analysis, and computational fluid dynamics (CFD) to assess vehicle performance. Each of these tools has limitations that preclude their exclusive use as the sole design and analysis method. Application of theoretical analysis is typically limited to very simple geometry and flight conditions with attached flow. Into the 21st century, the primary tool was experimental investigations. Application of CFD methods began when potential flow computations were applied in the 1960s and 1970s and provided some useful results. In the 1980s and 1990s computing costs decreased, and confidence in CFD increased. This has enabled CFD's use as the primary tool for analysis of aircraft cruise performance where there is well behaved, nonseparated flow over the vehicle. Even for the relatively well-behaved cruise conditions, there is still a need to use experimental investigations to confirm drag estimates and to identify conditions where there is flow separation. Experimental methods tend to be the dominant methods for transonic, high-lift, and powered-lift performance estimates.

Analysis of jet-induced effects has traditionally been based on experimental data, and several of their empirical correlations were shown in the previous chapters. In the 1970s several potential-flow, vortex-lattice, or panel-method based schemes[1–10] were developed to account for jet-induced interactions. A comparison[11] of five production surface-panel methods and one vortex-lattice method[12] demonstrated that good agreement for conventional aircraft aerodynamics can be achieved between any of these methods and experimental data. Several investigators[1–5,13–15] have extended potential flow-based methods to develop approximate predictive jet-in-a-crossflow (JICF) methods. One of the earliest and most complete calculation methods for jet-induced effects was developed by Wooler et al.[2,13] There were also applications[8–10,15–23] of potential flow-panel methods that represented major extensions of the Wooler computational method. However, these efforts provided only modest improvements in numerical solutions.

In the 1980s, the research emphasis shifted to the application of modern CFD methods to modular problems; such as the JICF, and to other experimental investigations specifically designed to obtain data for use in CFD verification. Several experiments[24–31] that used laser velocimetry to measure the JICF flowfield velocities in finer detail were conducted.[24–31] In the early 1990s there were some combinations of jet and freestream conditions where CFD was successfully applied. Specific examples will be presented for a jet in a crossflow, a jet in-ground effect, and for the complete Harrier aircraft in low-speed transition between hover and wing-borne flight. Finally some of the future trends in CFD will be discussed. These trends include application of hybrids of Reynolds-averaged Navier–Stokes (RANS) solutions with dense grids, large-eddy simulations (LES) for high-Reynolds-number separated flows, and the use of detached-eddy simulation (DES) for solution of other separated flows.

II. Panel Method Representation of Jet-Lift-Induced Effects

One of the earliest and most complete calculation methods for jet-induced effects was developed by Wooler et al.[2,13] in the 1960s and 1970s. This model used a double lifting line to represent a simple wing and a sink-doublet model to represent the jet. The Wooler method was improved later by incorporating a vortex lattice with the sink-doublet jet model to provide a more comprehensive, modular procedure.[32] For the first module, the incompressible jet model neglects viscous effects other than the entrainment caused by potential flow sinks or doublets. The entrainment of freestream flow into the jet and the pressure force on the jet boundary govern the two equations of motion of the jet. The entrainment parameters were obtained from experimental data such as Ricou and Spalding.[33] Yen[34] presents a complete discussion of the various entrainment parameters used by Wooler and other investigators. The jet cross section was varied from a circle at the exit to an ellipse as the jet is deflected by the freestream. The induced velocity flowfield caused by the jet is obtained from two singularities: 1) a uniform sink distribution on axes normal to the freestream at discrete locations along the jet centerline represent the entrained flow, and 2) a doublet distribution along the jet centerline represents the blockage and jet-induced circulation effects. Additional procedures were developed to represent nonaxisymmetric jets, jet pairs, and a jet in a nonuniform freestream.

A second module evaluates the jet-induced forces and moments using the vortex-lattice method[12] to represent lifting planforms. The method can give good agreement with experimental data for wing-body combinations by including the planform of the body in addition to the wing. Power-induced aerodynamic characteristics were evaluated by using a propulsion-induced camber distribution on the planform to satisfy the flow-tangency condition at the three-quarter chord of the horseshoe vortices of the lattice representation. This vortex-lattice program was used to improve the method of Ref. 2, which only represented the wing by a double lifting line.

Experimental data[35,36] were used to evaluate these methods. The model had a simple body shape and two wing-mounted nacelles mounted inboard on a unswept, tapered wing. Mineck[37] applied Wooler's method[32] and obtained good agreement with experimental data for the configuration with the nozzles deflected 90 deg in the aft position near the wing trailing edge (Fig. 7.1). For this case the jet-induced increase in wing lift (i.e., jet flap effect) dominated, and the flow on the wing was attached. In contrast, poor agreement was obtained with the nozzles deflected 90 deg in the forward position at $x/c = 0.11$. This result illustrated the inability of the method to account for separated flow on the wing lower surface aft of the jet. Other investigators with similar methods, such as Snel,[15] have found that the agreement is not good in the wake region.

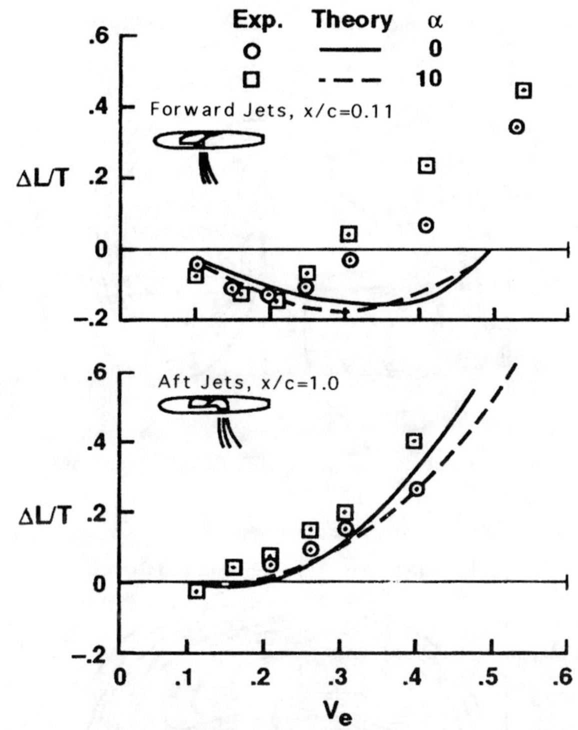

Fig. 7.1 Comparison between Wooler's method[32] and experimental data.[35]

For some cases, an empirical model of the separated wake improved the correlation of computed results with experimental data.

Adler and Baron[3] improved the Wooler-type method to predict the inner structure of the deflected jet more accurately. Two features of the jet were changed. First, the Chang[38] method for determining jet cross-section shape was incorporated, and second a method was developed to more adequately represent the velocity profile within the jet cross section. An entrainment model coupled with velocity decay and cross-section shape change and area growth were developed. The ideas of Keffer and Baines,[39] where the entrainment is composed of straight jet entrainment and vortical entrainment, were used. The vortical entrainment was represented by a vortex pair with local jet growth using a suggestion from Schwartz and Tulin.[40,41] An example where two jet cross sections are presented is given in Fig. 7.2 for two contrasting conditions of V_e and downstream distance. In both cases, the computed cross section (on the left) compared quite well with experimental results (on the right) from Kamotani and Greber.[42] Similar agreement was shown by Adler and Baron for conditions that include $0.083 < V_e < 0.25$ and $20 \text{ deg} < \delta j < 90 \text{ deg}$. Application of this method to pressure distributions on a flat plate or for aircraft configurations was not shown.

An empirical correlation method to predict the surface-pressure distribution was developed by Perkins and Mendenhall[8,16,43] for application to a JICF in a

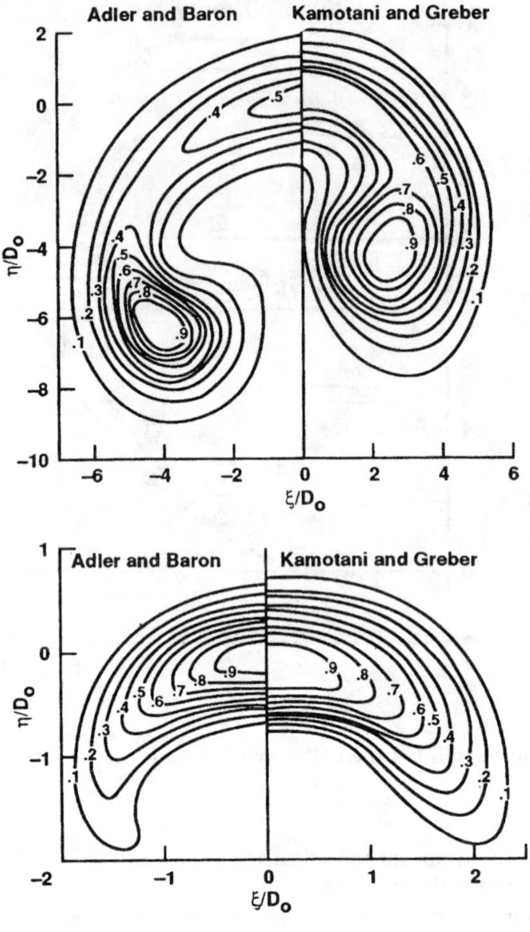

Fig. 7.2 Comparison between the Adler–Baron method[3] and experimental data[42] for two jet cross sections showing constant velocity contours normal to the met centerline path for $\delta_j = 90$ deg. a) Cross section located 23 diameters from exit and $V_e = 0.129$ b) Cross section located seven diameters from exit and $V_e = 0.26$.

flat plate or to a JICF in a body of revolution. The method includes a blockage model, an entrainment model, and viscous correction factors. For the flat plate, where there were adequate experimental data to define the needed correlation parameters, a reasonable comparison was demonstrated. For the body of revolution, there were inadequate data to properly define the correlation parameters.

In an attempt to improve the potential flow-based methods for application to actual aircraft configurations, panel methods were coupled with various higher-order models for the jet. An investigation by Maskew et al.[9] used the VSAERO panel method[44] with an iterative panel procedure for wake shape and entrainment for the jet and a source transpiration technique for boundary layers.

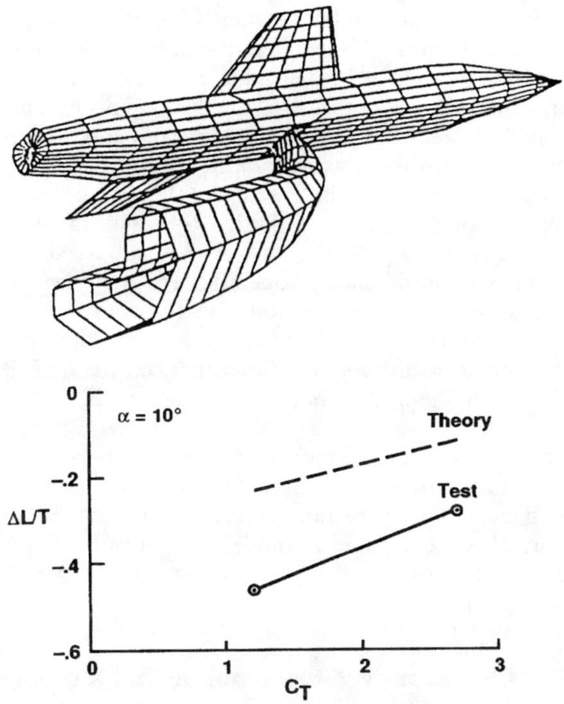

Fig. 7.3 Comparison of experimental data[48] and analysis of a V/STOL aircraft in transition flight using the PANAIR panel method[17] and a parabolized Navier–Stokes jet wake model.[20]

A separated wake model was attached vertically along the solid potential core of the jet at $\theta = 130$ deg. Limited results for the JICF flat-plate case were presented and were consistent with Wooler's results when a wake separation model was used. More complete comparisons with data[45] for a jet in a body of revolution were also presented. Here there is a discrepancy downstream of the jet because there was no separated wake model developed for this type of configuration.

In one example, Yoo and Strash[21] coupled VSAERO to represent the aircraft with a jet model using either ARC3D[46] or APPL,[47] a parabolized Navier–Stokes code developed by Roberts. Overlapping boundaries were used to couple the two methods. The model configuration[35,36] consisted of a simple wing body with a single lifting jet in the fuselage lower surface. The results were very sensitive to the radial grid density used to represent the jet.

A similar approach was taken by Howell.[19,20] The PANAIR panel method[17] with the Neumann boundary condition was used to represent the aircraft, and a parabolized Navier–Stokes solution with an entrainment boundary condition was used to represent the lifting jet. The jet shape and entrainment were determined using the Adler–Baron method[3] and unpublished data acquired by Adler. The method was applied to a simple high-wing aircraft configuration with a single lifting jet mounted in the fuselage.[48] A comparison between experiment and theory of jet-induced lift as a function of thrust coefficient is presented in Fig. 7.3 for the configuration at 10-deg angle of attack. The propulsion-induced lift loss is underestimated by about 20% of the thrust. This extension of previous methods is unable to account for the separated wake caused by the lifting jet.

A simple solution procedure was developed by Katz and Kern[22] for a delta wing with a jet blowing through a hole near the 0.5 root-chord location to represent an ejector lift concept. A vortex-lattice code with a separated rolled-up wake for the wing was developed. The jet was modeled by specifying a jet velocity normal to the wing in the ejector opening. The experimental model had a space around the jet where the separation effects could not physically occur. As a consequence, the problem of modeling a separation region is greatly reduced. The computed results for the thrust-induced lift and drag increments for a range of thrust coefficients at an angle of attack of 20 deg was surprisingly good.

To date, coupled-panel method/higher-order jet solutions have demonstrated reasonable comparisons with experimental data upstream and beside the jet exit. These methods have not demonstrated useful results downstream of the jet without an empirical wake separation model. As a result, they appear to offer no improvement over earlier potential flow aircraft/jet methods, and these approaches are rarely used anymore.

III. CFD Representation of the Jet in a Crossflow

The biggest advance since 1980 has come from the many CFD investigations[49–87] undertaken to obtain finite difference numerical solutions to the Navier–Stokes equations (NS) for the JICF. Several examples of recent results are examined with an emphasis on their comparisons with experimental data. It was shown by several investigators[50,70–72,75,78,82,84–86] that NS solutions can adequately represent most of the flowfield. The first example[86] of a Navier–Stokes equations finite difference solution for a turbulent jet is presented in Fig. 7.4 at an effective velocity ratio V_e of 0.25, where the velocities computed in the centerline plane are shown. This computation used a small (24 × 22 × 22) nonuniform grid (11,616 grid points) to represent the right half of the flowfield. The left half was accounted for by the use of a symmetry boundary condition. This symmetry approach has been used in most of the

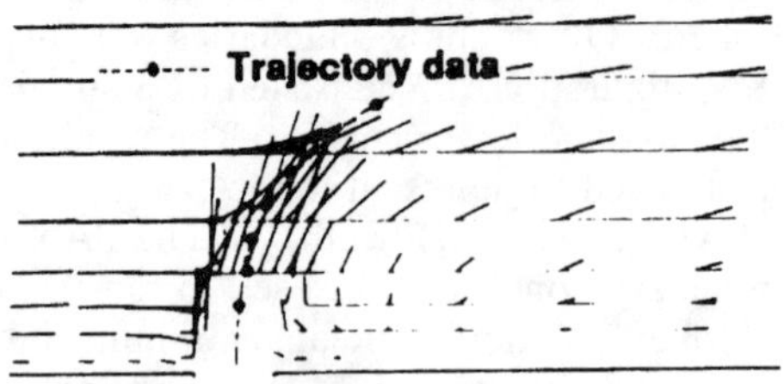

Fig. 7.4 Calculated[86] velocity vector plot for a JICF in the symmetry plane of jet for effective velocity ratio V_e = 0.25. Trajectory data are from Ref. 39.

computations presented next. The line at the bottom of the figure represents the surface where the jet is located and exited perpendicular to the freestream. The open gap in the line is the location of the jet exit. The velocity vectors show the penetration of the jet into the freestream flow and the subsequent defection of the jet downstream. The three-dimensional recirculating region upstream of the jet and the large wake region downstream of the jet are captured as well as the jet spreading and entrainment of surrounding fluid. The jet centerline path compares closely with Keffer and Baines[39] experimental data. Similar agreement with the jet path was shown in computations by Patankar et al.[85] However, the jet decay and wake region were not adequately represented by either computation.

In another NS computation, Oh and Schetz[75] used a penalty method finite element method (FEM), where the NS equations were solved using a generalized Galerkin technique. A simple eddy-viscosity turbulence model was used. The finite element mesh was small 30 × 12 × 12 (4320 grid points) and is not directly comparable with grid sizes used for finite difference methods. An example of the induced pressure distribution for an effective velocity ratio R of 4 is presented in Fig. 7.5 as a double "snail" plot. Experimental data[77] are shown on the left, and the FEM solution[77] is shown on the right. The upstream positive pressure regions compare closely. Beside the jet exit, the experimental negative pressure contours are larger in area than the FEM results. The largest difference is in the wake region. The FEM solution has a positive region, which is much like that obtained

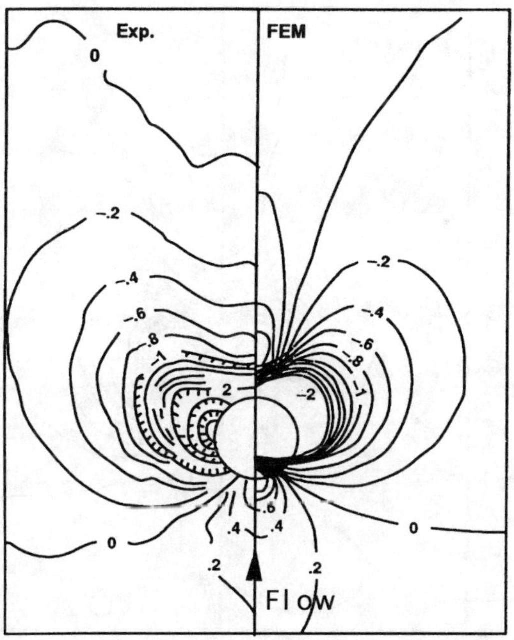

Fig. 7.5 Comparison of experimental[77] and Navier–Stokes computation[75] comparison of the pressure coefficients induced by a jet in a crossflow where $V_e = 0.25$.

from a simple potential flow solution. This is not a satisfactory solution for estimating induced effects on a V/STOL aircraft. In other FEM parabolized NS computations, Baker and colleagues[53–56] also obtained reasonable jet path and cross-section shape results but inadequate results in the wake region.

A NS computation by Kavsaoglu et al.[78] used a three-dimensional, compressible Navier–Stokes code, LANS3D.[79] The Baldwin–Lomax turbulence model[80] was used within five diameters of the flat plate, and the turbulence model of Oh and Schetz[75] was used in the jet region. A rectangular grid $59 \times 21 \times 45$ (55,755 grid points) was used for a circular jet computation. Along the symmetry plane the jet diameter was represented by 14 grid lines. The jet Mach number was 0.4. The nonuniform jet exit velocity profile was obtained from the characteristic boundary conditions (bc). The inflow x_{min} and outflow x_{max} used characteristic bc; the top boundary z_{max} used M_∞ of 0.1; and the bottom solid walls used the no-slip condition. The surface-pressure distribution obtained is compared in Fig. 7.6 with experimental data.[77] The negative pressure region beside the jet is represented well. Although the overall results are reasonably predicted, there are limitations. The computational results do not capture the positive pressure region ahead of the jet and the wake separation region.

The previous solutions, which used up to 60,000 grid points, found that the flows, especially near the jet exit plane and in the wake, were not adequately

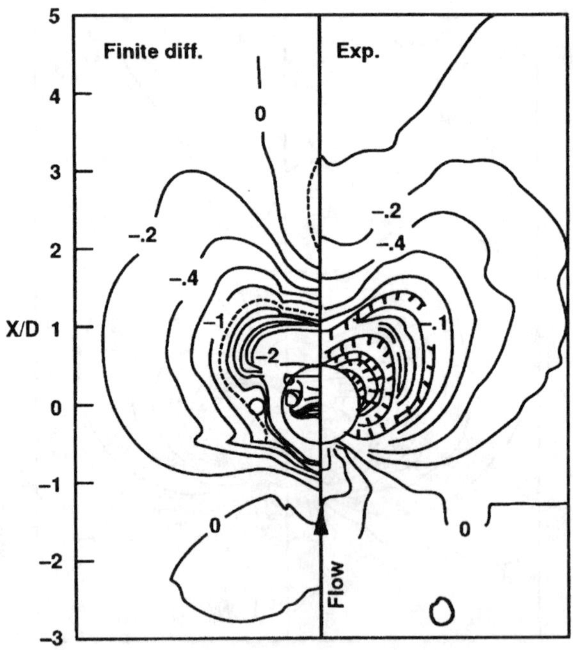

Fig. 7.6 Comparison of experimental[77] and Navier–Stokes computation[78] comparison of the pressure coefficients induced by a jet in a crossflow where $V_e = 0.25$.

resolved. Other computations have used up to 300,000 grid points in an attempt to improve the solutions near the jet exit and on the flat-plate surface. Roth[70] and Roth et al.[71,72] used a 55 × 55 × 50 (151,250 grid points) Cartesian grid. Chiu et al.[84] used the Chimera grid embedding technique[87] with two overlapping

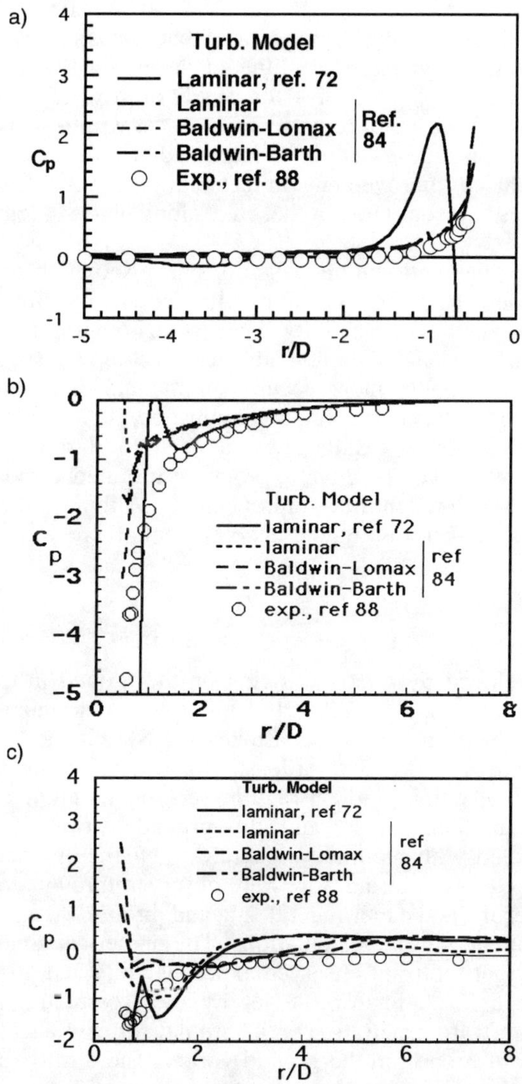

Fig. 7.7 Experimental[88] and Navier–Stokes computation[72,84] comparison of the pressure coefficients induced by a jet in a crossflow where $V_e = 0.167$. a) Upstream radial ray (0-deg azimuth) b) Spanwise radial ray (90-deg azimuth) c) Downstream radial ray (180-deg azimuth).

grids: 1) a 79 × 33 × 66 Cartesian grid (172,062 grid points) for the surrounding freestream and flat-plate boundary layer and 2) a 51 × 33 × 66 cylindrical grid (111,078 grid points) for the jet that was located along the jet vortex curve. The plate pressure coefficient distributions along three azimuthal rays (0, 90, and 180 deg) are presented in Fig. 7.7 for both computations and are compared with Fearn and Weston[88] experimental data. Along the zero deg ray upstream of the jet (Fig. 7.7a), the increased grid size used by Chiu et al.[84] greatly improves the agreement with experiment. Chiu showed that this is probably caused by the increased grid density near the edge of the jet. Along the 90- and 180-deg radial rays (Fig. 7.7b and 7.7c, respectively), the benefit of increased grid density is less apparent. Although there are differences near the jet within two diameters of the exit, there is little to suggest that either increased grid density or the turbulence models used produced improvements in the solutions. Particularly disappointing is the positive pressure computed in the wake along the 180-deg ray at large distances from the jet.

To date, these computational investigations have shown that the flow near the jet exit and on the surface downstream of the jet exit has not been adequately resolved. This is an area where activity is needed to both improve physical understanding of the flowfield and to exploit advances in computer hardware and algorithms, which could enable more accurate numerical solutions. At this time, comparisons with experimental data demonstrate that there is a need for improved solutions. These results show a need for larger grids (more than 300,000 grid points for a jet), more appropriate turbulence models both in and near the jet, or other unidentified improvements. Although there is activity in support of the Joint Strike Fighter development, there are few unrestricted reports available.

IV. Jet in-Ground Effect

The flow developed by a jet impinging on the ground in the presence of a crossflow, a jet in-ground effect (JIGE), was studied numerically by Van Dalsem[89] using the fortified Navier–Stokes (FNS) scheme.[90] An example of the results are shown in Fig. 7.8. At least 140,000 grid points were required to resolve the numerous high-gradient regions such as the ground boundary layer, the jet/freestream shear layer, and the ground vortex. This mixing was reasonably modeled with modified algebraic turbulence models. A number of interesting characteristics of the flow were observed through comparison of the computational results and experimental data and through the variation of the turbulent flow parameters in the calculations. The numerical simulation predicted the characteristic jet footprint observed experimentally and provided additional insight into the deformation of the jet by the freestream. For example, it appears that the forward penetration of the ground vortex is a strong inverse function of the level of mixing in the ground vortex. The results showed that more work will be required to accurately compute more sensitive low effective velocity ratio V_e flows.

Reference 91 presented a LES technique solution of a single impinging jet in crossflow and compared this both to a RANS-based k-ε turbulence model solution and to available experimental data. Illustrations were provided of the

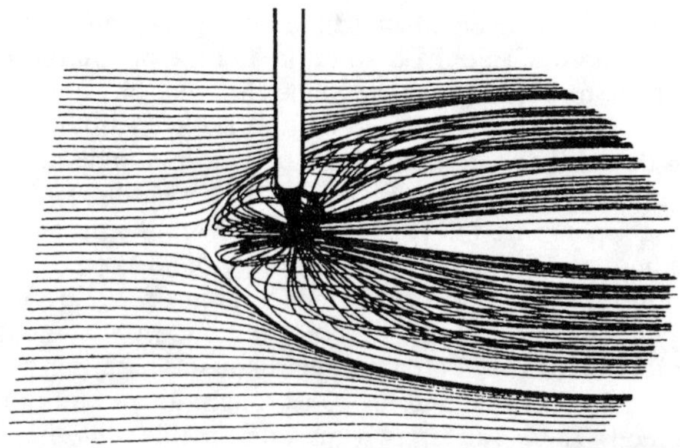

Fig. 7.8 Computational results[89] showing the perspective view of particle traces for a turbulent jet exiting from a nozzle in-ground effect with $V_e = 0.233$ and $h/D = 3$.

kind of time-resolved information describing the dynamics of turbulent structures that can be obtained from the LES approach, and it was pointed out that this knowledge would be essential in assessing hot-gas-ingestion (HGI) phenomena. In complex flow problems such as found in-ground-effect flows, the timescales in differing parts of the flowfield vary widely.

Figure 7.9 presents the streakline pattern obtained from a flowfield time averaged over many instantaneous solutions sampled from the LES predictions. The statistical averaging has been carried out over 100,000 time steps, or roughly three ground vortex turnover times; the statistics can therefore be

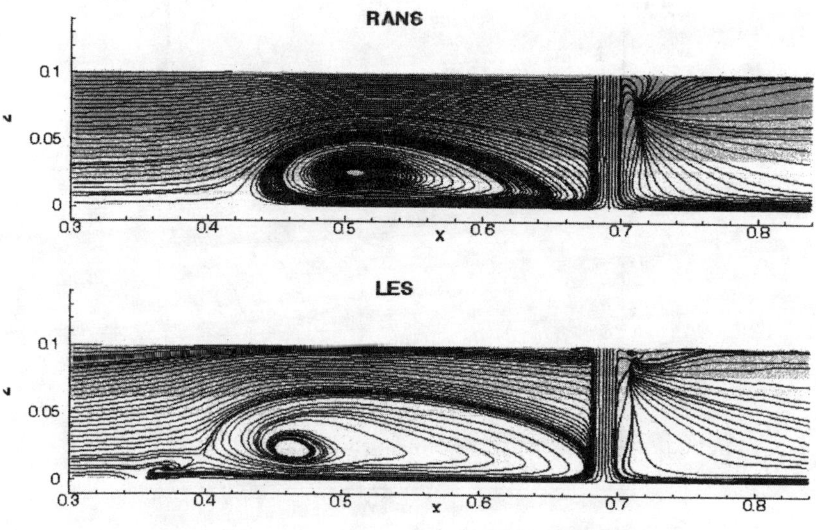

Fig. 7.9 Time-averaged streamline pattern.[91]

expected to be close to converged in a time-averaging sense. The figure shows also a comparison between the LES solution and the prediction using a RANS approach and a k-ε turbulence model. Some differences were evident between the two solutions. The shape of the ground vortex was different, being noticeably thicker in the LES solution, and having a more rounded rear shape where the vortex interacts with the entrainment into the forward edge of the jet. It was shown that these aspects are in better agreement with the measurements than in the RANS predictions. The axial extent of the ground vortex is also greater in the LES solution. The furthest forward penetration of the groundsheet flow is around 14.5 jet diameters in the RANS solution. In the measurements the leading edge of the vortex is at about 11 jet diameters. For this parameter therefore it seems that the LES method has not brought about an improvement over the RANS method. There can be two reasons for this. One is grid resolution. Whereas the k-ε turbulence model solution is almost certainly grid independent, no checks have yet been made in the LES solution for this. In particular, no effort has been made to examine whether the overall resolution of 600,000 cells is sufficient, or whether the mesh in the vicinity of the forward stagnation point is fine enough. The second possible explanation lies in the subgrid-scale (SGS) model. Correct prediction of the furthest forward penetration of the ground sheet against the crossflow will depend on the loss of momentum in the wall jet caused by skin friction. This might require both a finer grid in

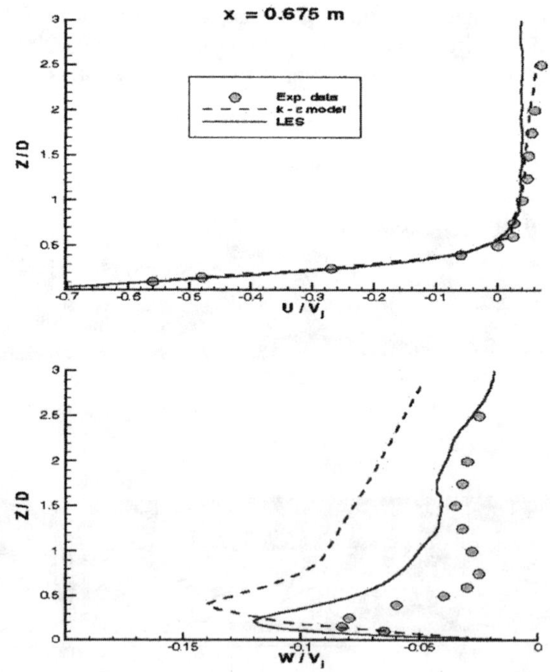

Fig. 7.10 Axial and vertical velocities near the jet edge.

the wall normal direction than currently used and also a more advanced SGS Reynolds-stress model.

The data in Fig. 7.10 examine the agreement of experimental data with the k-ε model and LES results for the two mean velocity components, streamwise u and vertical w, in the center axial plane at a cross section 0.625 diameter upstream of the forward edge of the jet. Both streamwise velocity solutions agree well with the experimental data. The LES vertical velocity solution shows a better representation of the vertical velocity component than the k-ε turbulence model solution. It is closer to the magnitude and shape of the experimental data.

Some of the other LES technique results from Ref. 91 were an improvement on what could be achieved using RANS-based CFD. This was particularly true in the vicinity of the jet impingement where comparison of the RANS solution with measured turbulent shear-stress values gave even the wrong sign in the k-ε turbulence solution, but showed excellent agreement in the LES prediction. Unfortunately, the LES simulation of flow events near the ground vortex leading edge is currently disappointing. The penetration distance was overpredicted and showed no improvement compared with a RANS prediction. Further study, involving checks of both mesh refinement and SGS model, is required to clarify the reasons for this in the LES prediction. In spite of this latter result, it is believed that the results of the study demonstrate that the LES approach has much promising potential in simulating ground-effect aerodynamics.

V. Analysis of a Complete Aircraft, the Harrier

A number of numerical investigations using simplified geometries have been carried out to validate computational methods and better understand the flow physics. Examples already presented include single and multiple jets in a crossflow.[92–94] Additional examples include delta wings with jet nozzles directed towards the ground and a RANS solution[95] for a simplified Harrier (wing, fuselage, inlets, and nozzles). All of these investigators cite two main problems in computing these flows: 1) the need for more accurate solution methods and 2) the need for a faster solution process. The need for a faster solution process is the key to improving the solution accuracy. One can hardly explore the use of different turbulence models and refined grids when a single solution can take many weeks.

Murman et al.[96,97] have focused on reducing the time to solution for the YAV-8B Harrier in-ground effect through process automation and exploitation of parallel computing. They began with the Harrier geometry reported in Ref. 95 and added a deflected flap and empennage for greater realism. They later included a dual time-stepping algorithm for improved time accuracy and solution robustness.[96] To date, more than 80 solutions have been carried out. Reference 98 described this process and progress made in reducing the time required to generate solutions.

Flow simulations of powered-lift vehicles, such as the YAV-8B Harrier in-ground effect, continue to be a challenge when using the RANS equations. A process is described in Ref. 99 that enables the generation of 35 time-dependent Navier–Stokes solutions for a YAV-8B Harrier aircraft in-ground effect in one week. The solution time is dramatically reduced through process automation and the use of large-scale parallel computers. A dual time-stepping algorithm that improves solution time accuracy and robustness is described. The static computations capture the suckdown and ground cushion effects. A method for computing the stability derivatives of lift variation with height and angle of attack is presented. Unsteady flow visualization and a discrete Fourier analysis are also used to identify and correlate key flow features with the time variation of lift.

Figure 7.11 shows that the mean lift coefficient C_L and dominant frequency f converge as the time step is decreased. The mean lift coefficient $C_{L,m}$ was obtained by integrating the surface pressures on the airframe and includes the aerodynamic lift and jet-induced lift. It does not include the engine thrust force. Reducing time step from the base solution ($\Delta t = 0.02$ with 15 subiterations) only produces a 1% change of the mean lift coefficient and a 6% change in the frequency, which is adequate for engineering purposes. Using smaller time steps or more subiterations is costly and was not considered justified. So it was decided to use these values to generate all 35 cases for the database. No comparison with experimental data was presented in this paper.

For the purpose of this book, a comparison with experimental data was attempted even though the flight condition is not consistent with the usual Harrier flight operation envelope. Some details of the configuration were not given in Ref. 99. The reference does give the nozzle boundary conditions for exit Mach number, velocity, and temperature. It is assumed that ambient atmospheric conditions can be used for the exit static conditions. Then the equation of state ($\rho = p/RT$) was used to compute the front and rear nozzle densities. These densities were then used to compute the front and rear jet dynamic pressures as 922 and 866 psf, respectively. Next it was assumed that the front and rear nozzle thrusts were equal, and then the combined equivalent jet dynamic pressure was computed as 894.3 psf. Finally for the $M = 0.15$ case, the V_e was computed as

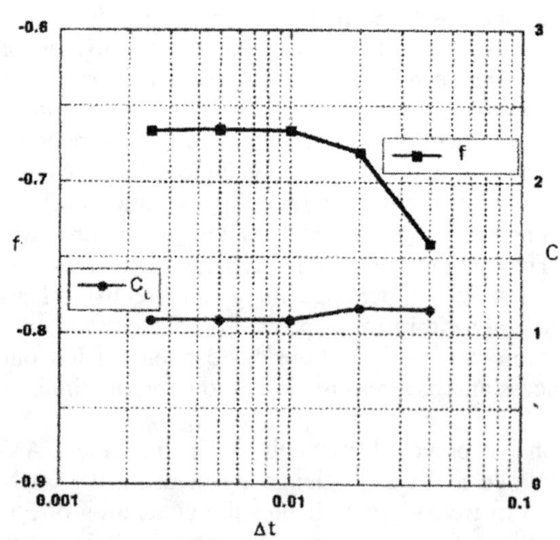

Fig. 7.11 Convergence of lift coefficient and dominant frequency with time step for $M = 0.05$ (33 kn) and $h = 15$ ft and $\alpha = 5$ deg.

0.06. Transition lift data are usually expressed as L/T and height as h/D_e. It was assumed that total thrust is 24,000 lbf and deflection is 81 deg, which are typical for a hovering Harrier. Using the AV-8B wing area as 230 ft^2, C_T is computed to be 28.3. Then because $L/T = C_{L,m}/C_T + \sin\ \delta_j$, L/T is computed as 0.96. Because thrust can be expressed as $F_G = 2A_j q_j$, the total jet exit area is computed as 13.4 ft^2, and D_e is then computed as 4.13 ft so that h/D_e is 3.63.

AV-8B data are available as trends but not readily available in specific numerical quantities, whereas data for the AV-8A prototype (Kestrel) are available without restriction in Ref. 100. The corresponding AV-8A experimental result is found to be $L/T = 0.95$ and is subject to refinement with more exact data for the AV-8B configuration. This result is good enough to suggest that there could be a good comparison between this computation and appropriate experimental data.

For example, Fig. 7.12 shows a streakline image from a RANS simulation of a Harrier in low-speed flight near the ground. A ground vortex is formed when the jet flows meet the oncoming ambient flow. The front jets on both sides of the aircraft also cause an upwash near the flow symmetry plane, resulting in a fountain vortex that forms in front of the ground vortex. These phenomena can pose difficult turbulence modeling issues. Moreover, low-speed freestream conditions, together with an unsteady flow with low frequencies (less than 1 Hz), contribute towards algorithm stiffness and very long compute times.

Figure 7.13 presents mean lift coefficient as a function of height and angle of attack for the entire database. The original 35 solutions are indicated by the solid symbols. An interpolation procedure, based on a local monotone cubic spline, is used to expand the computed database from 35 cases to over 2500 cases. This is accomplished by applying a one-dimensional interpolation operator successively in each parametric direction. The accuracy, of course, depends on having an adequate number of realistic CFD solutions. The present data show

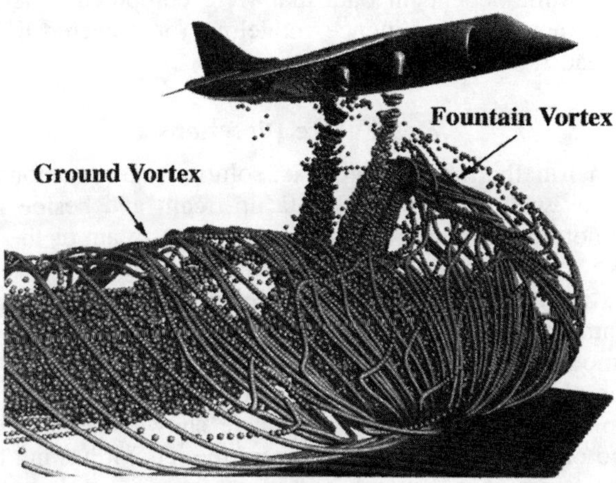

Fig. 7.12 Time-dependent streaklines where $M = 0.05$ (33 kn), $h = 30$ ft, and $\alpha = 9$ deg.

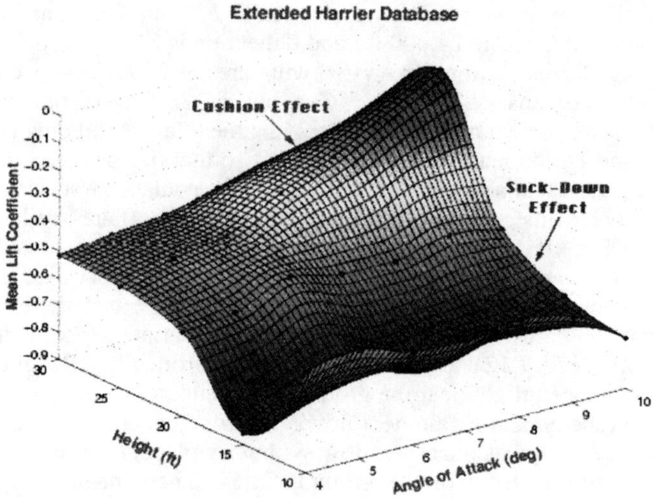

Fig. 7.13 Mean lift coefficient as a function of height and angle of attack for $M = 0.05$ (33 kn). The closed circles are the RANS solution, and the lines are the monotone cubic spline.

the combinations of height and angle of attack where there is a cushion effect as well as where there is a suck-down effect.

This effort[99] represents a successful start for CFD analysis of a complete V/STOL aircraft configuration. This research has reduced the computation time for high-fidelity time-dependent RANS solutions using improved computing technologies and state-of-the-art computing platforms. Unsteady flow visualization was generated using a Graphics Encapsulation Library to visualize the large amount of time-dependent data that were computed. The issues of grid refinement, accuracy, and turbulence modeling for powered-lift vehicles can now be more readily addressed.

VI. Future Directions

Coupled panel-method/higher-order jet solutions have demonstrated reasonable comparisons with experimental data upstream and beside a JICF. These methods have not demonstrated useful results downstream of the jet without an empirical wake separation model. As a result, they appear to offer no improvement over earlier potential flow aircraft/jet methods and are not useful for future development.

Currently modern CFD methods have been able to compute many of the mean flow characteristics of the JICF (experimental data such as Refs. 101–103). However, published CFD investigations have shown that the flow very near the jet exit and on the surface downstream of the jet exit has not been resolved. This is an area of current activity where both improved experimental understanding of the flowfield and advances in computer hardware and algorithms have enabled more elaborate numerical solutions. Comparisons with experimental data demonstrate that there is a need for improved solutions. These results show a need for larger grids (more than 300,000 grid points for a single jet), more appropriate turbulence models both in and near the jet, and other unidentified improvements. The rationale and need for continued research are demonstrated.

There is a need for high-quality, extensive experimental JICF data that will be suitable for the verification of current and future CFD results. Only limited laser velocimeter measurements have been made.[103] These results demonstrate the ability of this instrument to greatly improve the quality and precision of steady flowfield data. Both steady and unsteady measurements are needed. JICF needs to involve more complex geometry than a plain jet exiting normal to a flatplate surface. The effects of nozzle shape, coannular jets with different temperatures, exit louvers, and swirl typical of lift fans on the jet flowfield and on induced lift need to be investigated to support concepts currently being considered for the STOVL version of the JSF. There is also a need for experimental investigation of the effect of temperatures as high as those developed by modern high-pressure-ratio turbojet engines.

CFD investigation results have shown that more work will be required to accurately compute the JIGE for more sensitive low effective velocity ratio V_e flows. There is a need to experimentally quantify the unsteady character of the JIGE. In the case of two jets in-ground effect, experimental results show that there is not a true line of symmetry in the upflow region; instead, there is quite a lot of mixing and unsteady interaction between the upflow from the two jets. These results illustrate the importance of the modeling technique used in investigating these problems both experimentally and computationally.

A major CFD problem is the accurate calculation of separated flows. This has been a continuing deficiency. Recently a new approach called detached-eddy simulation (DES) has been proposed.[104] DES claims wide application, either in its initial form or in "cousins," which are defined as hybrids of RANS and LES, aimed at high-Reynolds-number separated flows. Several successful applications are referenced to provide confidence in this approach. Although grid generation is not easy for RANS calculations, the requirements of DES compound this difficulty. DES incorporates several turbulence treatments in the same field while being directed at complex geometry. The flowfield is divided into several regions such as an Euler region, a RANS region, and/or a LES region. The Euler region is upstream and to the sides and is never entered by turbulence or vorticity except if it is generated by shocks. This region extends to infinity and covers most of the volume but uses only a small share of the grid points. The RANS region is primarily the boundary layer. It is divided into a viscous region and an overlap region. The LES regions contain the vorticity and turbulence, at some point in the simulation, but are neither boundary layers, nor thin shear layers along which the grid can be aligned. The LES is divided into viscous, focus, and departure regions. There are intermediate zones between all regions. Regions are not distinguished by different equations being applied but by different priorities in the grid spacing. The time step is chosen for accuracy not stability. Although DES can produce inaccurate results if the grid is too coarse or the time step is too long, it rarely "blows up."

Opportunities for breakthroughs in large-scale computational simulation and design of aerospace vehicles are discussed in Ref. 105. The opportunities are obtained from speed-ups and robustness improvements in the underlying unit operations associated with simulation (geometry, modeling, grid generation, physical modeling, analysis, etc.). Further, an improved programming environment can synergistically integrate these unit operations to leverage the gains. The speed-ups result from reducing the problem setup time through geometry modeling and grid-generation operations and reducing the solution time through the operation counts associated with solving the discretized equations to a sufficient accuracy. Although the opportunities are addressed only at a general level,[105] there is an extensive list of references that contain further details. The goal is to enable greater inroads into the design process with large-scale simulations. It is expected that CFD work on development of the F-35 Joint Strike Fighter will achieve many of the needed improvements described earlier.

Nomenclature

A_j = jet exit area, ft^2
C_L = lift coefficient, $L/q_\infty S$
$C_{L,m}$ = mean lift coefficient obtain by integration of surface pressures includes both aerodynamic lift and jet-induced lift, $C_L + \Delta C_L$
C_p = pressure coefficient, p/q_∞
C_T = thrust coefficient, $FG/q_\infty S$
c = chord, ft
D, D_0 = diameter, ft
D_e = equivalent diameter, $\sqrt{4(\sum A_j/N)/\pi}$
f = frequency, 1/s
h = height, ft
M = Mach number
N = number of jets
p = pressure, psf
q_j = jet dynamic pressure, psf
q_∞ = freestream dynamic pressure, psf
R = gas constant, 1716.16 ft^2/s^2 °R
r = radius, ft
S = wing area, ft^2
T, F_G = thrust, lbf
V_e = effective velocity ratio, $\sqrt{q_\infty/q_j}$
V_j = jet velocity, fps
x, z = streamwise, vertical Cartesian directions
α = angle of attack, deg
ΔC_L = jet-induced lift coefficient
ΔL = jet-induced lift, lbf
Δt = time step, s
δ_j = jet deflection, deg
η, ξ = normal, lateral jet cross-section directions
ρ = density, lbf s^2/ft^4
θ = angular position on the jet cross section, where $\theta = 0$ deg faces the freestream

Subscripts

max = maximum
min = minimum

References

[1]Heltsley, F. L., and Parker, R. L., Jr., "Application of the Vortex Lattice Method to Represent a Jet Exhausting from a Flat Plate into a Crossflowing Stream," Arnold Engineering Development Center, AEDC-TR-73-57, Tullahoma, TN, June 1973.

[2]Wooler, P. T., Kao, H. C., Schwendemann, M. F., Wasson, H. R., and Ziegler, H., "V/STOL Aircraft Aerodynamic Prediction Methods Investigation," U.S. Air Force Flight Dynamics Lab., AFFDL-TR-72-26, Vol. I-IV, Dayton, OH, Jan. 1972.

[3]Adler, D., and Baron, A., "Prediction of a Three-Dimensional Circular Turbulent Jet in Crossflow," *AIAA Journal*, Vol. 17, No. 2, 1979; also *Proceedings V/STOL Aircraft Aerodynamics*, Naval Air Development Center, 1979, pp. 552–585.

[4]Beatty, T. D., "A Prediction Methodology for Propulsive Induced Forces and Moments in Transition and STOL Flight," *Proceedings NADC V/STOL Aircraft Aerodynamics*, Naval Air Development Command, Washington, DC, 1979, pp. 64–91.

[5]Siclari, M. J., Barche, J., and Migdal, D., "V/STOL Aircraft Prediction Technique Development for Jet-Induced Effects," NAPTC Rept. PDR-623-18, Naval Air Development Command, Washington, DC, April 1975.

[6]Siclari, M., Migdal, D., and Palcza, J. L., "Development of Theoretical Models for Jet Induced Effects on V/STOL Aircraft," *Proceedings Prediction Methods for Jet V/STOL Propulsion Aerodynamics*, Naval Air Development Command, Washington, DC, 1975, pp. 998–1015, (avail. from DCC as AD-A024-023).

[7]Foley, W. H., and Sansome, J. A., "V/STOL Propulsion-Induced Aerodynamic Hover Calculation Method," Rept. NADC-TR-78242-60, National Research Lab, The Netherlands, Feb. 1980.

[8]Perkins, S. C., and Mendenhall, M. R., "Computer Programs to Predict Induced Effects of Jets Exhausting into a Crossflow," NASA CR-166,591, June 1984.

[9]Maskew, B., Strash, D., Nathman, J., and Dvorak, F. A., "Investigation to Advance Prediction Techniques of the Low-Speed Aerodynamics of V/STOL Aircraft," NASA CR-166,479, Feb. 1983.

[10]Furlong, K. L., and Fearn, R. L., "A Lifting Surface Computer Code with Jet-in-Crossflow Interference Effects; Vol. I—Theoretical Description; Vol. II—Users Guide for WBWJAS," NASA CR-166,524, Aug. 1983.

[11]Margason, R. J., Kjelgaard, S. O., Sellers, W. L., III, Morris, C. E. K., Jr., Walkey, K. B., and Shields, E. W., "Subsonic Panel Methods—A Comparison of Several Production Codes," AIAA Paper 85-0280, Jan. 1985.

[12]Margason, R. J., and Lamar, J. E., "Vortex-Lattice FORTRAN Program for Estimating Subsonic Aerodynamic Characteristics of Complex Planforms," NASA TN D-6142, Feb. 1971.

[13]Wooler, P. T., "Development of an Analytical Model for the Flow of a Jet into a Subsonic Crosswind," NASA SP-218, Sept. 1969, pp. 101–119.

[14]Heltsley, F. L., and Kroeger, R. A., "A General Jet Efflux Simulation Model," NASA SP-218, Sept. 1969, pp. 165–180.

[15]Snel, H., "A Model for the Calculation of the Properties of a Jet in a Cross Flow," NLR TR 74080 U, June 1974.

[16]Perkins, S. C., and Mendenhall, M. R., "A Correlation Method to Predict the Surface Pressure Distribution on an Infinite Plate or a Body of Revolution from Which a Jet Is Issuing," NASA CR-152,345, May 1980.

[17]Magnus, A. E., and Epton, M. E., "PAN AIR—A Computer Program for Predicting Subsonic or Supersonic Linear Potential Flows about Arbitrary Configurations Using a Higher-Order Panel Method," NASA CR-3251, April, 1980.

[18]Roberts, D. W., "A Zonal Method for Modeling 3-D Aircraft Flow Fields with Jet Plume Effects," AIAA Paper No. 87-1436, June 1987.

[19]Howell, G. A., and Snyder, L. D., "Development of V/STOL Methodology Based on a Higher Order Panel Method," NASA CR 166491, May 1984.

[20]Howell, G. A., "Automated Surface and Plume Simulation Procedure for Use with Aerodynamic Panel Codes," NASA CR-177420, May 1986.

[21]Yoo, S., and Strash, D. J., "Zonal Approach to V/STOL Aerodynamics," *Journal of Aircraft*, Vol. 27, No. 10, 1990, pp. 866–872.

[22]Katz, J., and Kern, D., "Effect of Vertical-Ejector Jet on the Aerodynamics of Delta Wings," *Journal of Aircraft*, Vol. 27, No. 5, 1990, pp. 408–412.

[23]Fearn, R. L., "Progress Towards a Model to Describe Jet/Aerodynamic-Surface Interference Effects," *AIAA Journal*, Vol. 22, No. 6, 1984, pp. 752, 753.

[24]McLaughlin, D. K., "Laser Doppler Velocimeter Measurements in a Turbulent Jet Exiting into a Cross Flow," Arnold Engineering Development Center, AEDC-TR-71-262, Tullahoma, TN, Jan. 1972.

[25]Rudinger, G., and Moon, L. F., "Laser-Doppler Measurements in a Subsonic Jet Injected into a Subsonic Cross Flow," *Journal of Fluids Engineering*, Vol. 43, Sept. 1976, pp. 516–520.

[26]Aoyagi, K., and Snyder, P. K., "Experimental Investigation of a Jet Inclined to a Subsonic Crossflow," AIAA Paper 81-2610, Dec. 1981.

[27]Fearn, R. L., Doddington, H., and Westphal, R., "LDV Studies of a Jet in a Crossflow," Report NADC-80238-60, Naval Air Development Command, Washington, DC, Sept. 1981.

[28]Snyder, P., and Orloff, K. L., "Three-Dimensional Laser Doppler Anemometer Measurements of a Jet in a Crossflow," NASA TM-85997, May 1984.

[29]Orloff, K. L., and Snyder, P. K., "Using a Three-Dimensional Laser Anenometer to Determine Mean Streamline Patterns in a Turbulent Flow," NASA TM 85,948, July 1984.

[30]Snyder, P. K., and Orloff, K. L., "Three-Dimensional Laser Doppler Anemometer Measurements of a Jet in a Crossflow," NASA TM 85,997, Sept. 1984.

[31]Sislian, J. P., and Cusworth, R. A., "Laser Doppler Velocimeter Measurements of Mean Velocity and Turbulent Stress Tensor Components in a Free Isothermal Swirling Jet," Univ. of Toronto, UTIAS Rept. 281, Ontario, March 1984.

[32]Wooler, P. T., "Propulsion-Induced Effects on a Supersonic V/STOL Fighter/Attack Aircraft," *Proceedings NADC V/STOL Aircraft Aerodynamics*, Naval Air Development Command, Washington, DC, 1979, pp. 173–190.

[33]Ricou, F. P., and Spaulding, D. B., "Measurements of Entrainment by Axisymmetrical Turbulent Jets," *Journal of Fluid Mechanics*, Vol. II, Pt. 1, Aug. 1961, pp. 21–32.

[34]Yen, K. T., "The Aerodynamics of a Jet in a Crossflow," Rept. NADC-78291-60, Dec. 1978.

[35]Mineck, R. E., and Schwendemann, M. F., "Aerodynamic Characteristics of a Vectored-Thrust V/STOL Fighter in the Transition-Speed Range," NASA TN D-7191, May 1973.

[36]Mineck, R. E., and Margason, R. J., "Pressure Distribution on a Vectored-Thrust V/STOL Fighter in the Transition-Speed Range," NASA TM X-2867, March 1974.

[37]Mineck, R. E., "Comparison of Theoretical and Experimental Interference Effects on a Jet VTOL Airplane Model," *Proceedings Prediction Methods for Jet V/STOL Propulsion Aerodynamics*, Naval Air Development Command, Washington, DC, 1975, pp. 1016–1034 (avail. from DCC as AD-A024-023).

[38]Chang, H.-C., "Aufrollung Eines Zylindrischen Strahles Durch Querwind," Ph.D. Dissertation, Univ. of Gottingen, Germany 1942; also "The Roll-up of a Cylindrical Jet in a Cross Flow," translated and edited by K.S. Nagaraja, and H. O. Schrade, USAF-ARL 73-0131, University of Gottingen, Gottingen, Germany, Sept. 1973 (in English).

[39]Keffer, J. F., and Baines, W. D., "The Round Turbulent Jet in a Cross Wind," *Journal of Fluid Mechanics*, Vol. 15, Pt. 4, 1963, pp. 481–496.

[40]Tulin, M. P., and Scwartz, J., "The Motion of Turbulent Vortex Pairs in Homogeneous and Density Stratified Media," Hydronautics, Inc. Tech. Rept. 231–15, Naval Air Development Command, Washington, DC, 1971.

[41]Schwartz, J., and Tulin, M. P., "Chimney Plumes in Natural and Stable Surrounding," *Atmospheric Environment*, Vol. 6, No. 1, 1972, pp. 19–35.

[42]Kamotani, Y., and Greber, I., "Experiments on a Turbulent Jet in a Cross Flow," *AIAA Journal*, Vol. 10, No. 11, 1972, pp. 1425–1429.

[43]Perkins, S. C., and Mendenhall, M. R., "A Study of Real Jet Effects on the Surface Pressure Distribution Induced by a Jet in a Crossflow," NASA CR-166,150, March 1981.

[44]Maskew, B., "Prediction of Subsonic Aerodynamic Characteristics: A Case for Low-Order Panel Methods," *Journal of Aircraft*, Vol. 19, No. 2, 1982, pp. 157–163.

[45]Aoyagi, K., and Snyder, P. K., "Experimental Investigation of a Jet Inclined to a Subsonic Cross Flow," AIAA Paper 81-2610, Dec. 1981.

[46]Pulliam, T. H., and Steger, J. L., "Implicit Finite Difference Simulations of Three-Dimensional Compressible Flow," *AIAA Journal*, Vol. 18, No. 2, 1980, pp. 159–167.

[47]Roberts, D. W., "Prediction of Subsonic Aircraft Flows with Jet Exhaust Interactions," AGARD, CP-301, May 1981.

[48]Vogler, R. D., "Interference Effects of Single and Multiple Round or Slotted Jets on a VTOL Model in Transition," NASA TN D-2380, July 1964.

[49]Jones, W. P., and Mc Guirk, J. J., "Computations of a Round Jet Discharging into a Confined Crossflow," *Turbulent Shear Flows 2*, Springer-Verlag, Heidelburg, Germany, 1980, pp. 233–245.

[50]White, A. J., "The Prediction of the Flow and Heat Transfer in the Vicinity of the Jet in Cross Flow," American Society of Mechanical Engineers, ASME-80-WA/HT-26, Jan. 1980.

[51]Demuren, A. O., "Numerical Calculations of Steady Three-Dimensional Turbulent Jets in Cross Flow," *Computational Methods in Applied Mechanical Engineering*, Vol. 37, May 1983, pp. 309–328.

[52]Demuren, A. O., "Modeling Turbulent Jets in Crossflow," *Encyclopedia of Fluid Mechanics*, Vol. 2, Gulf Publishing Co., Houston, TX, 1985, pp. 430–465.

[53]Baker, A. J., Orzechowski, J. A., and Manhardt, P. D., "A Numerical Three-Dimensional Turbulent Simulation of a Subsonic Interaction VSTOL Jet in Crossflow Using a Finite Element Algorithm," Rept. NADC 79021-60, Naval Air Development Command, Washington, DC, July 1980.

[54]Baker, A. J., Orzechowski, J. A., Manhardt, P. D., and Yen, K. T., "A Three-Dimensional Finite Element Algorithm for Prediction of V/STOL Jet-Induced Flows," AGARD CP-308, Paper 26, Nov. 1981.

[55]Baker, A. J., and Orzechowski, J. A., "An Assessment of Factors Affecting Prediction of Near-Field Development of a Subsonic VSTOL Jet in Crossflow," NADC Tech. Rept. 81177-60, Naval Air Development Command, Washington, DC, Sept. 1982.

[56]Baker, A. J., Snyder, P. K., and Orzechowski, J. A., "Three Dimensional near Field Characterization of a VSTOL Jet in Turbulent Crossflow," AIAA Paper 87-0051, Jan. 1987.

[57]Claus, R. W., "Numerical Calculation of Subsonic Jets in Crossflow with Reduced Numerical Diffusion," AIAA Paper 85-1441, 1985.

[58]Claus, R. W., and Vanka, S. P., "Multigrid Calculations of a Jet in Crossflow," AIAA Paper 90-0444, Jan. 1990.

[59]Pulliam, T. H., and Steger, J. L., "Implicit Finite Difference Simulations of Three-Dimensional Compressible Flow," *AIAA Journal*, Vol. 18, No. 2, 1980, pp. 159–167.

[60]Roberts, D. W., "Prediction of Subsonic Aircraft Flows with Jet Exhaust Interactions," AGARD, CP-301, May 1981.

[61]Roberts, D. W., "A Zonal Method for Modeling Powered-Lift Aircraft Flow Fields," NASA Contract NAS2-12801, Amtec Engineering, Inc., AEI-T88100.01, NASA Ames Research Center, Moffett Field, CA, Dec. 1988.

[62]Roberts, D. W., "A Zonal Method for Modeling Powered-Lift Aircraft Flow Fields," NASA Contract NAS2-13357, Amtec Engineering, Inc., AEI-T93100.01, NASA Ames Research Center, Moffett Field, CA, Sept. 1991.

[63]Mongia, H. C., Reynolds, R. S., and Srinivasan, R., "Multidimensional Gas Turbine Combustion Modelling Applications and Limitations," *AIAA Journal*, Vol. 24, No. 6, 1986, pp. 890–904.

[64]Holdeman, J. D., and Srinivasan, R., "Modeling Dilution Jet Flowfields," *Journal of Propulsion and Power*, Vol. 2, No. 1, 1986, pp. 4–10.

[65]Karki, K. C., and Mongia, H. C., "Recent Developments in Computational Combustion Dynamics," AIAA Paper 89-2808, 1989.

[66]Sykes, R. I., Lewellen, W. S., and Parker, S. F., "On the Vorticity Dynamics of a Turbulent Jet in a Crossflow," *Journal of Fluid Mechanics*, Vol. 168, July 1986, pp. 393–412.

[67]Needham, D. J., Riley, N., and Smith, J. H. B., "A Jet in Crossflow," *Journal of Fluid Mechanics*, Vol. 188, Jan. 1988, pp. 159–184.

[68]Harloff, G. J., and Lytle, J. K., "Three-Dimensional Viscous Flow Computationss of a Circular Jet in a Subsonic and Supersonic Cross Flow," AIAA Paper 88-3703, June 1988.

[69]VanOverbeke, T. J., and Holdeman, J. D., "A Numerical Study of the Hot Gas Environment Around a STOVL Aircraft in Ground Proximity," AIAA Paper 88-2882, Aug. 1988.

[70]Roth, K. R., "Numerical Simulation of a Subsonic Jet in a Crossflow," Society of Automotive Engineers, Paper 872343, Dec. 1987.

[71]Roth, K. R., "Influence of the Thin-Layer Approximation on Jet in Crossflow Computations," AIAA Paper 90-3056, Aug. 1990.

[72]Roth, K. R., Fearn, R. L., and Thakur, S. S., "Evaluation of a Navier-Stokes Prediction of a Jet in a Crossflow," *Journal of Aircraft*, Vol. 29, No. 2, 1992, pp. 185–193.

[73]Shang, J. S., McMaster, D. L., Scaggs, N., and Buck, M., "Interaction of Jet in Hypersonic Cross Stream," *AIAA Journal*, Vol. 27, No. 3, 1989, pp. 157–163.

[74]Huang, G.-P., "Model and Computation of Three-Dimensional Turbulent Jets in a Crossflow," Ph.D. Dissertation, Ecole Centrale de Lyon, France, June 1989.

[75]Oh, T. S., and Schetz, J. A., "Finite Element Simulation of Complex Jets in a Crossflow for V/STOL Applications," *Journal of Aircraft*, Vol. 27, No. 5, 1990, pp. 389–399.

[76]Shi, Z., Wu, J. M., and Wu, J. Z., "Symmetric and Asymmetric Jets in a Uniform Crossflow," AIAA Paper 91-0722, Jan. 1991.

[77]Kavsaoglu, M. S., and Schetz, J. A., "Effects of Swirl and High Turbulence on a Jet in a Crossflow," *Journal of Aircraft*, Vol. 26, No. 6, 1989, pp. 157–163.

[78]Kavsaoglu, M. S., Akmandor, I. S., Ciray, S., and Fujii, K., "Navier-Stokes Simulation of Two and Three Dimensional Jets in Crossflow," AIAA Paper 91-1743, June 1991.

[79]Fujii, K., and Obayashi, S., "Practical Applications of New LU-ADI Scheme for the Three-Dimensional Navier-Stokes Computation of Transonic Viscous Flows," AIAA Paper 86-0513, Jan. 1986.

[80]Baldwin, B. S., and Lomax, H., "Thin Layer Approximation and Algebraic Model for Separated Turbulent Flows," AIAA Paper 78-257, Jan. 1978.

[81]Bruiatskii, E. V., and Kuz'menko, G., "Calculation of the Cross-Sectional Shape of a Jet in a Cross Flow," Inst. Gidromekhaniki, Kiev, Ukraine, Gidromekhanika, No. 63, 1991, pp. 15–29.

[82]Demuren, A. O., "Characteristics of 3D Turbulent Jets in Crossflow," NASA TM 104337, April 1991.

[83]Kim, S.-W., and Benson, T. J., "Fluid-Flow of a Row of Jets in Crossflow," AIAA Paper 92-0534, Jan. 1992.

[84]Chiu, S. H., Roth, K. R., Margason, R. J., and Tso, J., "A Numerical Investigation on a Subsonic Jet in Crossflow," AIAA Paper 93-0870, Jan. 1993.

[85]Patankar, S. V., Basu, D. K., and Alpay, S. A., "Prediction of the Three-Dimensional Velocity Field of a Deflected Turbulent Jet," *Journal of Fluids Engineering*, Vol. 99, No. 4, Dec. 1977, pp. 758–762.

[86]Chien, C. J., and Schetz, J. A., "Numerical Solution of the Three-Dimensional Navier-Stokes Equations with Application to Channel Flows and a Bouyant Jet in a Cross-Flow," *Journal of Applied Mechanics*, Vol. 42, 1975, pp. 575–579.

[87]Benek, J. A., Buning, P. G., and Steger, J. L., "A 3-D Chimera Grid Embedding Technique," AIAA Paper 85-1523, Jan. 1985.

[88]Fearn, R. L., and Weston, R. P., "Induced Pressure Distribution of a Jet in a Crossflow," NASA TN D-7916, July 1975.

[89]Van Dalsem, W. R., "Study of V/STOL Flows Using the Fortified Navier-Stokes Scheme," *Computational Fluid Dynamics*, edited by G. de Vahl Davis and C. Fletcher, Elsevier Science, North Holland, 1988, pp. 725–735.

[90]Van Dalsem, W. R., and Steger, J. L., "Using the Boundary-Layer Equations in Three-Dimensional Viscous Flow Simulation," AGARD-CP-412, Paper 24, Oct. 1986.

[91]Tang, G., Yang, Z., Page, G. J., and McGuirk, J. J., "Simulation of an Impinging Jet in Crossflow using an LES Method," AIAA Paper 2002-5959, Nov. 2002.

[92]Van Dalsem, W. R., Panaras, A. G., and Steger, J. L., "Numerical Investigation of a Jet in Ground Effect with a Crossflow," Society of Automotive Engineers, Paper 872344, Dec. 1987.

[93]Barata, J., Durao, D., and McGuirk, J., "Numerical Study of Single Impinging Jets Through a Crossflow," AIAA Paper 89-0449, Jan. 1989.

[94]Barata, J., "Numerical and Experimental Study of Fountain Flows Produced by Multijet Impingement on a Ground Plane," AIAA Paper 91-1806, June 1991.

[95]Smith, M. H., Chawla, K., and Van Dalsem, W. R., "Numerical Simulation of a Complete STOVL Aircraft in Ground Effect," AIAA 91-3293-CP, Sept. 1991.

[96]Murman, S. M., Chaderjian, N. M., and Pandya, S.A., "Automation of a Navier-Stokes S&C Database Generation for the Harrier in Ground Effect," AIAA Paper 2002-0259, Jan. 2002.

[97]Chaderjian, N. M., Pandya, S., Ahmad, J., and Murman, S. M., "Parametric Time-Dependent Navier-Stokes Computations for a YAV-8B Harrier in Ground Effect," AIAA Paper 2002-0950, Jan. 2002.

[98]Pandya, S., Chaderjian, N., and Ahmad, J., "Parametric Study of a YAV-8B Harrier in Ground Effect Using Time-Dependent Navier-Stokes Computations," AIAA Paper 2002-3056, Jan. 2002.

[99]Chaderjian, N., Pandya, S., Ahmad, J., and Murman, S. M., "Progress Toward Generation of a Navier-Stokes Database for a Harrier in Ground Effect," AIAA Paper 2002-5966, Nov. 2002.

[100]Margason, R. J., Vogler, R. D., and Winston, M. M., "Wind-Tunnel Investigation at Low Speeds of a Model of the Kestrel (XV-6A) Vectored-Thrust V/STOL Airplane," NASA TN D-6826, July 1972.

[101]Kamotani, Y., and Greber, I., "Experiments on a Turbulent Jet in a Cross Flow," NASA CR-72893, June 1971.

[102]Fearn, R. L., and Weston, R. P., "Induced Pressure Distribution of a Jet in a Crossflow," NASA TN D-7916, July 1975.

[103]Dennis, R. F., Tso, J., and Margason, R. J., "Induced Surface Pressure Distribution of a Subsonic Jet in a Crossflow," AIAA Paper 93-4861, Dec. 1993.

[104]Spalart, P. R., "Young-Person's Guide to Detached-Eddy Simulation Grids," NASA CR-2001-211032, July 2001.

[105]Alexandrov, N. et al., "Opportunities for Breakthroughs in Large-Scale Computational Simulation and Design," NASA TM-2002-211747, June 2002.

Index

Supporting Materials

A complete listing of titles in Progress in Astronautics and Aeronautics and other AIAA publications is available at http://www.aiaa.org.